*Pharmacological Methods
in Phytotherapy Research*

Volume I

Selection, Preparation and Pharmacological Evaluation of Plant Material

We would like to thank the British Council who have supported the Link Programme between the School of Pharmacy and the University of Ibadan

Pharmacological Methods
in Phytotherapy Research

Volume I

Selection, Preparation and Pharmacological Evaluation of Plant Material

Elizabeth M. Williamson
The School of Pharmacy, University of London, UK
David T. Okpako
Faculty of Pharmacy, University of Ibadan, Nigeria
Fred J. Evans
The School of Pharmacy, University of London, UK

JOHN WILEY & SONS
Chichester · New York · Brisbane · Toronto · Singapore

Other Wiley Editorial Offices

John Wiley & Sons, Inc., 605 Third Avenue,
New York, NY 10158-0012, USA

Jacaranda Wiley Ltd, 33 Park Road, Milton,
Queensland 4064, Australia

John Wiley & Sons (Canada) Ltd, 22 Worcester Road,
Rexdale, Ontario M9W 1L1, Canada

John Wiley & Sons (Asia) Pte Ltd, 2 Clementi Loop #02-01
Jin Xing Distripark, Singapore 0512

Library of Congress Cataloging-in-Publication Data

Williamson, Elizabeth M.
    Pharmacological methods in phytotherapy research / Elizabeth M.
Williamson, David T. Okpako, Fred J. Evans.
        p. cm.
    Includes bibliographical references and index.
    ISBN 0 471 94216 2 (alk. paper).—ISBN 0 471 94217 0 (alk. paper).
    1. Materia medica, Vegetable—Research—Methodology.
2. Pharmacognosy. I. Okpako, D.T. II. Evans, Fred J., 1943–
III. Title.
RS164.W55 1996
615′.32′072—dc20   95-46352
                                                         CIP

British Library Cataloguing in Publication Data

A catalogue record for this book is available from the British Library

ISBN 0 471 94216 2 (Hard cover)
ISBN 0 471 94217 0 (Paperback)

Typeset in 10/12pt Photina by Acorn Bookwork, Salisbury, Wiltshire.
Printed and bound in Great Britain by Biddles Ltd, Guildford.
This book is printed on acid-free paper responsibly manufactured from sustainable forestation, for
which at least two trees are planted for each one used for paper production.

Contents

Introduction

The exploitation of plants for medicines has a long and honourable history, since at one time all drugs were obtained from natural sources. It was also linked with the initial development of the science of pharmacology, which used natural products to elucidate physiological processes and even define them, hence the naming of 'nicotinic' and 'muscarinic' receptors, and more recently 'endorphins' from 'endogenous morphines'. Other plant compounds of historical importance which are still used today include the cholinesterase inhibitor physostigmine, the anticholinergic atropine, adrenergic neurone depleters like reserpine and the antimitotic alkaloid colchicine. This story, rather than drawing to a close, is now experiencing a renaissance, led partly by the successful development of new compounds and partly by a great public interest in all things natural, ecologically sustainable and biodegradable.

Newer pharmacological tools, such as colforsin (an adenyl cyclase stimulator), ginkgolide B (a specific PAF antagonist) and the phorbol esters which activate protein kinase C, are at the forefront of biochemical research and all are obtainable only from plant sources.

The pharmacological evaluation of natural products therefore forms an intrinsic part of pharmacognosy, one of the original disciplines of the pharmaceutical profession. Although in the UK the subject matter of pharmacognosy has largely been dissipated, in other parts of the world it remains intact, and covers all aspects of the science of safe and professional use of natural products. Herbal medicine is currently enjoying a revival in popularity in the West and is of course the primary form of medicine in many parts of the world. Traditional Chinese, Ayurvedic and Unani systems of medicines are spreading throughout the world with increasing population movement. Herbal medicines, as effective and potent medicines, require evaluation by standard scientific methods in order to be used to their full effect.

Natural products are the basis of many standard drugs used in modern medicine, and are so widely used that many laymen and even some members of the medical profession are unaware that they are of plant origin. Obvious examples are digoxin, hyoscine (scopolamine), theophylline, ergometrine, ephedrine, pilocarpine, vincristine and vinblastine.

Plant compounds have also served as templates for the development of many drugs, including the local anaesthetics such as lignocaine from cocaine; analgesics and cough suppressants developed from opiates; neuromuscular blockers from tubocurarine; carbenoxolone from glycyrrhetinic acid; chloroquine from quinine; etoposide from podophyllotoxin; cromoglycate from khellin; bromocriptine from the ergot alkaloids; and many more.

Some of the more important new developments are taxol, from the Pacific Yew, *Taxus brevifolia*, recently licensed for the treatment of ovarian cancers, and artemisinin, an antimalarial from *Artemisia annua*. Garlic, an ancient remedy used in Egypt in the times of the Pharoahs, is only now showing its true potential as an antisclerotic and antithrombotic; and ginger is increasingly being used as an anti-emetic, especially in morning sickness of pregnancy and for motion sickness. The latter two highlight the use of foodstuffs as medicines or, viewed conversely, the pharmacology of foods occurring in our regular diet. Even if the potency of some of these compounds is not sufficient to justify their development as pharmaceuticals, their effects on the body must be considered because of the sheer volume of their ingestion. An example of this is the oestrogenic isoflavonoids found in soya and other legumes.

Plant drugs also supply efficient treatments for some conditions where 'conventional' medicine has little to offer, such as liver damage, where lignans from *Silybum marianum* can prevent fatalities induced by death cap mushroom (*Amanita phalloides*) poisoning and infectious hepatitis, and immune stimulants such as the polysaccharides from the coneflower, *Echinacea*, are used to treat viral infections. Those plant materials used as drugs of abuse, including the notorious *Cannabis sativa*, are still, like the opiates, yielding novel and useful compounds.

Plant products which are used as pharmaceutical excipients need to have either an inert pharmacological profile or an advantageous one. For example, the alginates obtained from seaweed have a natural carminative action on the stomach as well as a thickening effect for liquid formulations and are often found in antacid preparations. Natural colourings like riboflavin, the carotenoids and anthocyanins are used for food products and medicines and have benefits such as vitamin or antioxidant activity. These substances must be evaluated in the same way as other medicines, and the methods given in this book are as applicable to them as they are in the search for new drugs.

Traditionally, evaluation of natural drugs consisted of botanical and chemical identification and standardisation. The modern discipline of pharmacognosy encompasses these studies and extends to the search for new biologically active compounds from natural sources. This is largely but not exclusively concentrated on plants and strategies for selecting these are discussed in Chapter 1.

Isolation of active compounds must be monitored at each stage with biological tests, as there is no point in carrying out lengthy and expensive extraction procedures to end up with a useless compound. Such biological tests should be simple and economical, while giving as much information as possible, since numerous tests will be performed. Most pharmacological techniques can be used for isolated pure compounds, but some are too sophisticated to use on crude extracts or fractions. This book is a selection of some of the most commonly used methods for biological testing (or screening) applied to natural products, giving illustrations of results from actual publications and references to other tests, theoretical background and recent work. They can be used to produce a bioassay if a suitable calibration curve can be established, as shown in Chapter 2.

Some of the methods given are classical experimental pharmacology; they are reproduced here since the older practical pharmacology books are now difficult to obtain. Other methods, for example measuring effects of plant drugs on liver damage, platelet aggregation and cholesterol levels, are given in more depth because they have not been previously collected together in a practical textbook. A summary of chemical extraction methods and references is given in Chapter 3, and the remaining parts of the book are devoted to pharmacological methods arranged in therapeutic categories.

Volume II will deal with toxicological evaluation, chemotherapy, isolated cell and enzyme systems, and alternatives to animal testing.

1

The Use of Plant Remedies in Indigenous Medical Systems

Introduction

During the last few decades, there has been a resurgence of interest in plants as sources of medicines and of novel molecules for use in the elucidation of physiological/biochemical phenomena. There are a number of reasons for this. First, there is the genuine expectation in developing countries that their drug problems can be alleviated through a sensible scientific exploitation of medicinal plants, some of which have been used for generations by indigenous populations. Then there is the world-wide 'green' revolution which is reflected in the belief that herbal remedies are safer and less damaging to the human body than synthetic drugs. Furthermore, underlying this upsurge of interest in plants, is the fact that many important drugs in use today were derived from plants or from starting molecules of plant origin: digoxin/digitoxin, the vinca alkaloids, reserpine and tubocurarine are some important examples. Plants have also yielded molecules which are extremely valuable tools in the characterisation of enzymes and the classification of receptor systems: physostigmine, morphine, muscarine, atropine, nicotine and tubocurarine are important examples. Some scientists thus expect that the plant kingdom holds the key to the understanding of complex human biochemistry/pathology and the cure of man's perplexing diseases. The initial optimism, engendered by the idea that a sophisticated understanding of receptor systems and of the biochemistry of disease would pave the way to predictable drug development, has not been realised. Therefore, laboratories around the world are engaged in the screening of plants for biological activity with therapeutic potential. One major criterion for the selection of a plant for such study is traditional healers' claims for its therapeutic usefulness. It is thus worth reflecting on the cultural environment in which tradi-

1

tional healers use plant remedies, as well as the methods of plant use, in order to strengthen the research design.

Causes of Illness and the Use of Plant Remedies in Indigenous Systems of Medicine

Virtually every human society evolved an indigenous health care system to cope with illness. In Western technologically advanced societies, traditional prescience notions of the causes of disease and how to manage it have given way to modern ideas based on scientific biomedical theories. In the less technologically developed societies, traditional modes of thought still dominate the forms of medical practice seen in those societies. It is imperative that we do not ignore the thought processes behind these systems for two reasons: first, it is the continued use of plants for the treatment of disease in these systems that has invigorated our interest in phytotherapy; second, the experience crystallised in cultural practices of medicine can often be of value in the biomedical scientist's search for understanding of complex aspects of healing.

There is a great variation between traditional societies in their perception of the causes of serious illness, and the extent to which these beliefs are articulated. For example, there are ancient records on traditional Chinese medicine, the Indian Ayurveddic and Pakistani Unani systems. The theories of these Asian systems have been refined and elaborated over several millenniums. The fundamental concepts of Chinese medicine are embedded in Confucianism and Taoism which by 600 BC stood as two fully evolved philosophies (Chow, 1984). These bodies of knowledge have been systematised so that, for example, colleges of Ayurveddic medicine now exist in parts of the world, and acupuncture, practised in traditional Chinese medicine, is now an accepted method in biomedicine. On the other hand, the African therapeutic systems, though just as ancient, have remained more informal, less organised, based as they are on oral traditions (Okpako, 1991).

In spite of the different levels of articulation, indigenous traditional systems share certain common attributes. The most important of these is the tendency to see man as an integral part of nature and to regard a harmonious relationship with the rest of nature as being essential for good health. Most traditional medical systems therefore emphasise the *holistic* nature of their approach to the management of illness. They tend to be concerned, not with specific *diseases* but with the state of *illness* which is believed to be brought about by an imbalance, a disharmony in the elements that govern the integrity of the individual in his/her particular cultural environment. In the Chinese system, this idea is highly developed in the doctrine that illness is caused by an imbalance in the elaborate opposites, *yin* and *yang*. Treatment is aimed at restoring the body to a state of harmony. In the African perception, the spiritual component of the human existence is greatly emphasised. Illness is believed to be due to disharmony between the sufferer, his spiritual (ancestors and gods) and social worlds. To re-establish harmony by means of suitable sacrifices and ritual exposure of hidden guilt is therefore a major objective in the treatment of serious illness.

The second attribute shared by traditional systems of medicine is the use of plant remedies (*herbology*). Again the Asian pharmacopoeias of plant remedies are well worked and developed over a period of several millenniums. In these systems, the use of plants in the treatment of illness is in the context of beliefs as to the cause of disease. Therefore, plants are selected not so much on the basis of their chemical constituents, but on the basis of their perceived ability to restore harmony, and often informally, according to the *doctrine of signatures*. Relatively little is known of the thought processes underlying the use of plants in traditional African medicine. All one can say with some certainty is that the use of plants in this system is also within the context of the cultural beliefs as to what causes illness. Therefore, if one is to undertake a phytotherapeutic analysis of the plants used in traditional medicine, one must work with a consciousness of the culture in which the remedies are used.

The following analysis of the underlying considerations in the use of plant remedies in traditional African medicine has been given for this purpose. The theory has been constructed from observations of contemporary practice by traditional African healers, but much of it could apply in general terms to other indigenous health care systems.

Indigenous African Medicine

Traditional medical systems usually categorise illnesses into two broad groups:

1. Those illnesses which can be treated without religious invocations through divination and for which most adults in the community would know a remedy and use it, as it were, without a prescription. Aches and pains, minor injuries such as cuts and bruises are examples. Serious injuries such as fractures and bullet wounds may also be treated by specialists with or without divination.
2. In general, any illness serious enough to threaten the life of the patient, whether it be chronic illness whose cause is not immediately obvious or serious accident/injury, is usually thought to have a supernatural underpinning. Treatment thus involves divination by specialists to find out what gods or ancestor spirits have been offended and what sacrifices need to be made to appease the offended supernatural entity.

On the grounds of both incompatibility with introduced monotheistic religious beliefs and biomedical scientific theories, supernatural explanations of illness are viewed with scepticism by medical scientists. It must nevertheless be appreciated that it is within the context of such psychosomatic views

of illness that traditional healers employ herbology. From a careful examination of the ways in which plant preparations are used in traditional African
medicine, the following observations can be made:

1. Plant remedies for minor ailments owe their therapeutic benefits to
    physicochemical properties. For example, fresh leaves squeezed to stop
    bleeding contain tannins or other haemostatic principle. Similarly plant
    remedies used in the treatment of fevers contain antipyretic principles
    (e.g. *Azadirachta indica*, *Morinda lucida*). In such instances, where the
    clinical symptoms are self-evident, treatment is clearly directed at alle-
    viating the physical component of the illness.
2. It is different when plant preparations are used in the treatment of
    serious chronic illness—swelling of extremities (e.g. due to congestive
    heart failure), chronic persistent debilitating cough (as in pneumonia),
    fits (as in epilepsy), unexplained lumps and painful episodes (as in
    cancer) or paralysis (as in stroke); in these instances the traditional
    healer would not have the necessary technology for accurate diagnosis,
    and the underlying cause may simply be attributed to supernatural inter-
    vention or witchcraft. In such cases the plant remedy is part of a total
    ritual treatment regimen. We may presume, therefore, that the plant
    preparation is used not so much for its pharmacological properties as for
    its ritualistic (placebo?) significance.

There are several lines of argument that can be used to support this view.
For one thing, plant preparations are frequently used in a manner which
does not permit a pharmacodynamic or pharmacokinetic interpretation of
their efficacy: the plant preparation meant to treat an internal ailment may
simply be encapsulated and worn around the waist or placed under the
pillow or sleeping mat; the preparation is administered without regard for
dosage, which would not be the case if the plant was used for the pharmacological properties of its constituents. More than anything else, the issue of
dosage offers an insight into the role of plant remedies in the treatment of
illness in African traditional systems.

The Pharmacology of Indigenous African Medicine

In African medicine, the preparation is usually administered with only a
casual reference to the quantities. The patient is advised to take a tumblerful,
a calabashful or a seashellful of a decoction from time to time; such instruc-
tion has more to do with *how* the medicine should be used rather than the
quantity to be taken. If the decoction was perceived as a pharmacological tool
to be administered in strictly regulated doses, then the weight of plant
material, volume of solvent and the time interval between doses would need
to be determined. But precise measurement techniques which enable the
modern technologist to determine weight, volume and time in absolute units

do not exist in most traditional African cultures. Therefore, the decoction is made with a handful of leaves/bark in some water or alcohol.

We can make the assertion that in the evolution of the pharmacopoeia of traditional remedies, poisonous plants which proved harmful when taken without strict adherence to dosage were excluded from the common treatment regimen. Poisonous plants—arrow poisons, fish poisons, ordeal poisons, poisonous mushrooms—which are major sources of drugs used in modern medicine, are known to, but not usually employed by, traditional healers for therapeutic purposes. Rather, plants which are frequently used as medicines are also food components and spices.

These observations give us important insight into what can be described as the fundamental assumptions underlying the treatment of illness in traditional African medicine. Indigenous healers and their clients associate serious illness with an antisocial act committed in secret, a sin. Since ancestor spirits are believed by the people to be custodians of group morality, the fear of supernatural reprisal constitutes a stress which may gnaw at the individual and render him/her more susceptible to disease than a 'clean' member of the community. A major goal in therapy is, therefore, to bring the secret act into the open (through divination and confessions, carried out by specialists) for ritual treatment. The plant preparation is used as part of the ritual; the combined effects of the different components of the ritual is to remove the stress and allow the patient's own natural body defence mechanisms to surge. In this system the medicine is not directed primarily at killing or preventing the growth of an infective agent, or at correcting a specific biochemical lesion. This is to be contrasted with the principle of selective toxicity in modern pharmacology, which relies on high technology for accurate dosage of poisons.

The absence of strict rules of dosage should therefore be seen as an inherent attribute of indigenous African medicine. It enabled a pharmacopoeia of safe medicines (safe in the context of traditional dosage forms and usage) to evolve. Therefore, the fact that traditional healers do not adhere strictly to dosage, should not be used simply as grounds for denigrating traditional systems of medicine, for example by assuming that it is a reflection of the traditional healers' ignorance, or that it means that traditional medicines are unsafe.

Criteria for Selecting Plants for Investigation

The number of species of higher plants on this planet is estimated to be between 370 000 and 500 000. All higher plants elaborate chemical secondary metabolites that are of potential medicinal interest. Therefore, the determination of the criteria for selecting plants for phytotherapeutic investigation is perhaps as important an exercise as the investigation itself. The following selection criteria are suggested as a guide:

1. *Selection based on traditional usage.* This is a popular basis for selecting plants for investigation, especially in societies where traditional medicine of some sort is a major form of health care. The reasons are obvious. The investigation is easily justified, a clear objective is pursued through identification of plant material and there is a suitable biological model. However, in using this method, it is well to bear in mind the preceding argument regarding the perceived role of herbal remedies in disease management in the cultural setting under consideration. If a traditional healer claims success in the treatment of a particular disease, the medical scientist working from the above selection criteria would expect to find, in the plant extract, a chemical constituent with an appropriate pharmacological activity. This expectation is based on the principle of selective toxicity which is applicable in modern medicine but not in traditional systems, as we have argued. The extract of the plant selected on this basis should therefore be screened not just for biological activity appropriate to the claims of the healer, but also on as wide a range of other models as possible. This suggestion is based on the logic of the argument as well as past experience. The Madagascar rose periwinkle, which was eventually found to contain the powerful anticancer agents, the vinca alkaloids, was reputed to be a cure for diabetes mellitus in traditional medicine. Also, the extract of the roots of *Fagara xanthoxyloides* (*orinata* in Yoruba) was originally investigated for antimicrobial activity, consistent with its use as chewing stick (chewed to a brush and used to clean the teeth), and the observation that its users tended to be free of dental caries; but in laboratory tests the extract serendipitiously showed antisickling activity. This then sparked off much research into *Fagara* for antisickling potential, for which the plant is famous today. Furthermore, in view of recent findings that some common constituents of plants (polysaccharides and saponins, for example) are strong immunomodulators, tests for this activity should be included in extracts of plants used in traditional medicine.

2. *Poisonous plants.* Poisonous plants are known to indigenous populations, but because these are not usually used as medicines by traditional healers, medical scientists searching for drugs in plants do not often include them as a group for phytochemical investigation. A search for highly specific and potent compounds that can be used as drugs in modern medicine or as probes for the elucidation of biological phenomena is likely to be more productive among poisonous plants than in plants used regularly by traditional healers. Many of the most important drugs of plant origin used in medicine today come from poisonous plants, e.g. tubocurarine (arrow poison); atropine (poison); picrotoxin (fish poison); muscarine (poisonous mushrooms); dicoumarol (poisonous clover); physostigmine (ordeal poison). Other compounds of current medicinal interest obtained from plants are a series of tumour-promoting and

pro-inflammatory phorbol esters (from poisonous members of the Order Euphorbiales). The approach in screening extracts of poisonous plants would be to test the extract in a random battery of biological models in search of an interesting activity. But the test can be more focused if the nature of the poisoning caused by the plant is known, e.g. muscle paralysis. Although it is not always possible to foresee the therapeutic benefit of this kind of investigation, it is justifiable on scientific grounds or even as explanation for the mechanism of the poisoning and a possible way to counteract it.

3. *Selection based on chemical composition.* The laboratory may decide, for example because of facilities available to it, to extract a certain class of compounds, such as alkaloids, for investigation. Then different species of plants which are known to contain alkaloids, whether or not these have been used in traditional medicine, are extracted and screened on as wide a range of models as possible. This approach is greatly helped by chemo-taxonomic information relating different classes of compounds to different plant species.

4. *Screening for a specific biological activity.* Another way to proceed is to decide on a set of pharmacological laboratory models for a disease and test extracts of plants selected according to any criteria on the models for possible therapeutic usefulness in the treatment of the disease. The models may be designed to search for anticancer, antihypertensive, anti-inflammatory or cholesterol-lowering activity, etc. On the basis of com-puterised databanks, it is possible to say that one species of plant is more likely to yield the activity of interest than another. In the absence of a databank, plants can be screened at random. Results from such proce-dures suggest that for anticancer activity, random selection is no less productive than selection based on traditional claims. Poisonous plants were more active than plants selected on either of the above criteria.

5. *Combination of criteria.* The decision as to what plants to investigate can also be based on a combination of criteria. For example, plants used in traditional medicine and which are also known to contain particular types of compounds, e.g. alkaloids or glycosides, may be investigated. This approach would depend on the available chemical expertise and facilities; the extract should nevertheless be screened for as wide a range of biological activities as possible.

Conclusion

Phytotherapy research is expensive, in terms of both materials and expertise. It cannot be overemphasised, therefore, that the choice of what to investigate should be made after careful consideration. The criteria listed above should serve only as a guide. The important point for those working in parts of the world where traditional medicine is practised extensively, is this: although traditional usage is a useful guide, it is not sufficient to build the entire protocol on that alone. Unfortunately, many investigators fall into the temptation to do so, with disappointing results.

FURTHER READING

Chow, E.P.Y. (1984). Traditional Chinese medicine: a holistic system. In *Alternative Medicines. Popular and Policy Perspectives*, ed. J.W. Salmon. London: Tavistock, pp. 114–137.
Okpako, D.T. (1991). *Principles of Pharmacology. A Tropical Approach*. Cambridge University Press.

2

Presentation of Results

The Dose–Response Curve (Figures 2.1, 2.2)

The idea that the response of a cell, organ or the whole organism to a drug is proportional to the dose of the drug is of fundamental importance in drug action. Pharmacological investigations, therefore, nearly always entail the measurement of dose–response relationships. The basic assumption is that the drug molecule binds to a specific site on the cell, generally referred to as the receptor site. There is a finite number of such sites on the cell; the elementary theory is that the response elicited by the drug is related to the proportion of the receptor sites occupied by the drug molecules. The total number of receptor sites is known, but it is assumed that the fraction of receptor sites occupied by the drug is directly related to the dose of drug administered. The response is therefore a function of the dose of drug. Therefore, in practice, what is measured is the response to a given dose of drug. A plot of the dose (usually on a log scale) against the response gives a sigmoid curve; the latter is more useful for pharmacological analysis than a hyperbola, which is obtained if the dose is plotted on a linear scale.

There are some important points to note about a log dose–response curve:

1. The response increases by small amounts at first after the threshold, until it reaches about 20% of the maximum response; then the same increases in dose give rise to relatively large increases in response. Beyond about 80%, again increases in dose give rise to only small changes in response until the maximum response, when an increase in dose no longer causes an increase in response.
2. The portion of the curve lying between about 20% and 80% of the maximum is virtually a straight line, which is useful for quantitative analysis of drug action, e.g. comparing potencies of two drugs.
3. The maximum response reflects the proportion of the receptor sites occupied as well as such factors as the 'efficacy' of the drug in initiating a response after combining with the receptor, and the dissociation rate

constant of the molecule at the receptor. It is important to establish the maximum in all measurements involving dose–response relationships in order to be certain that the range of doses used includes submaximal doses. It is apparent from (1) above that doses causing below 20% and above 80% of the maximum response cannot be very useful in quantitative drug analysis without other mathematical transformation.

Some Applications of the Dose–Response Curve

Determination of Relative Potencies

The dose–response curve can be used to measure the relative potencies of drug or extract. From the curve the following parameters, which are measures of potency, can be determined, e.g. *EC50*, the concentration required to produce 50% of the maximum response: this is *IC50* if the response is inhibition. Another related quantity is PD_2, the molar concentration of agonist producing 50% of the maximum response. Agonist affinity constants can also be found from dose–response curves.

The Nature of the Interaction Between Drugs or Between Drug and Plant Extract, e.g. Potentiation or Inhibition

In the presence of a potentiating drug, the dose–response curve is shifted to the left of control. Inhibition of response can occur as a result of competitive block of agonist receptors (the lock and key principle), or by some other mechanism that may or may not involve receptor occupation. In competitive inhibition, the maximum response is unchanged, i.e. the two lines are parallel, provided the concentration of the antagonist is not excessively high. Competitive antagonism is also called *surmountable* antagonism because the agonist response can be restored in the presence of the antagonist by increasing the concentration of the agonist.

In non-competitive antagonism, the threshold dose of agonist is not markedly increased, but the maximum response is depressed in the presence of the antagonist. Crude plant extracts often produce this kind of effect in *in vitro* models. The effect arises when the blocking action takes place, as it were, beyond the receptor, such as effects on intracellular second messengers—adenylate cyclase, protein kinase C, ion channels, etc.

The Potency of a Competitive Antagonist and Receptor Classification

A quantity called the pA_2 can be determined from dose–response curves. The pA_2 is defined as the molar concentration of a *competitive* antagonist at which the dose of the agonist acting at the same receptor has to be doubled in order to produce the effect that it did before the addition of the antagonist. The pA_2

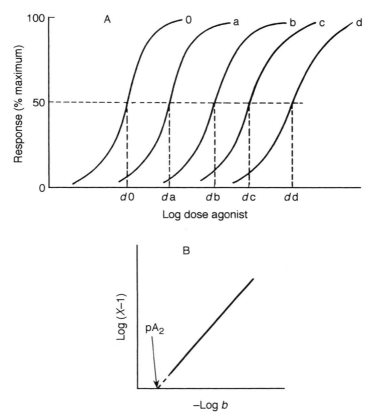

Figure 2.1 Schild Plots. (A) Log dose–response curves for agonist alone (0) and in the presence of increasing concentrations of antagonist (a, b, c, d). The dose ratios (x) at 50% maximum response for each antagonist concentration are d_a/d_o, d_b/d_o etc. (B) The plot of log $(x-1)$ versus the negative log of the molar concentration of antagonist ($-\log b$) corresponding to the x value, gives a straight line with intercept on the $-\log b$ axis $= pA_2$ of the antagonist and a slope $= 1$. From *Principles of Pharmacology: A tropical approach*, D.T. Okpako (1991), Cambridge University Press, Cambridge, UK, with permission.

is determined from a *Schild plot* (Arunlakshana and Schild, 1959). The higher the pA_2 value, the more potent is the antagonist at the particular receptor. For example, the pA_2 for mepyramine at histamine H_1-receptors is of the order of 9, whereas that for atropine at the same receptors is about 5. This means that mepyramine is about four orders of magnitude (about 10 000 times) more potent than atropine as a competitive antagonist at histamine H_1-receptors. At muscarinic receptors, the order of potency is reversed by roughly the same extent. The pA_2 can also be used to define a receptor type. If the receptor mediating the action of an agonist in two or more different

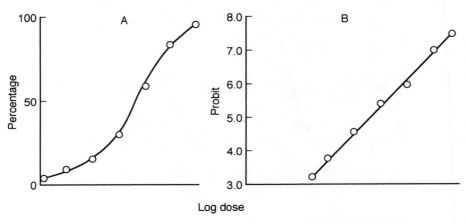

Log dose

Figure 2.2 A hypothetical log–dose quantal response curve with responses plotted (A) on a linear scale and (B) on a probit scale. From *Principles of Pharmacology: A Tropical Approach*, D.T. Okpako (1991), Cambridge University Press, Cambridge, UK, with permission.

tissues is the same, then the pA_2 values for the antagonist should be the same when determined in the two tissues, provided the agonist used acts on these same receptors. If the pA_2 values differ significantly, then the agonist is activating different types of receptors in the tissues/organs under investigation.

pA_2 determinations are best done in isolated preparations. *In vivo* determinations are liable to be complicated by metabolic and pharmacokinetic factors.

Therapeutic Index (TI)

The therapeutic index is a measure of the margin of safety of a drug. It is expressed numerically as a ratio of the dose of drug causing death or a harmful effect (lethality) in a proportion (x) of the sample, to the dose causing the required therapeutic effect (effective dose) in the same or higher proportion (y) of the sample. Thus TI can be expressed as LD50/ED50, where the number 50 refers to 50% of the population. A higher proportion may be used for the therapeutic dose, e.g. LD5/ED95 to increase confidence that the drug will be safe. Dose–response curves are used for this determination.

Types of Dose–Response Curves

The dose–response curves referred to above are usually constructed from *graded responses*, i.e. the curve is made from increasing responses in the same organ or tissue as the dose of drug is increased. Another type of curve is made from what are called *quantal responses*. In this case, the response is an end point such as death or convulsion of the animal; the animal either responds or it does not. Such measurements are used in the subjective measurement of pain in such models as the *hot-plate method*. If the response is graded as 1 (or 100%) and no response as 0 (0%), then, when groups of animals are given graded doses of drug or extract, the number responding in a group can be expressed as a percentage. A sigmoid curve is obtained when the response is plotted against log dose. The usual practice, based on statistical theory, is to transform the curve into a straight line by probit transformation.

There are published tables showing percentages and their corresponding probit values.

FURTHER READING

Arunlakshana, O. and Schild, H.O. (1959). Some uses of drug antagonists. *Br. J. Pharmacol. Chemother.*, **14**, 48–59.

3

Preparation of Plant Material

Plant material consists of a complex mixture of cells and chemical substances, and although in herbal medicine it may be used in this crude form, perhaps compressed into a tablet, for most purposes an extract will be needed; certainly for biological testing this will be the case. Extractions range from a simple infusion, or herbal tea, to a purified single chemical entity. For potent materials a pure compound, or at least a biologically standardised preparation, will need to be isolated, since accidental variations in dose would be dangerous. This is why, for example, a valuable drug like digoxin has a relatively recent history of use whereas its source, the foxglove, is not a traditional herbal medicine.

It is not always possible to isolate a single active principle, e.g. if it is very unstable, unknown, or if a range of compounds are responsible for the biological activity. In this case a bioassay can be developed using a suitably purified extract. Recent advances in chromatographic techniques mean that isolation of compounds that were previously found to be too unstable can now be achieved, and modern methods of structure elucidation can be carried out on very small amounts of material.

Correct botanical identification of plant material is essential, followed by preparation for extraction. Storage of damp plant material will lead to spoilage by fungi and bacteria, or fermentation. Drying is the usual method of preservation but may not always be possible if denaturing of the active compounds occurs, e.g. fresh ginger contains gingerols which are degraded to the more pungent shogaols during dehydration. Drying with heat is unsuitable for volatile constituents and freeze-drying (lyophilisation) is usually safer. Alcohol may be used as a preservative and solvent; however, it is flammable and bulky. Deep-freezing is the safest way of protecting labile compounds although it is not always convenient to keep large quantities of plant material in the freezer. There is no single perfect method and so the advantages and disadvantages of each must be evaluated.

In this chapter the various methods of extraction will be considered, followed by simple tests for different classes of compounds (alkaloids, glycosides, saponins, etc.) and a summary of chromatographic separation techniques and solubilisation.

Extraction of Chemicals from Plants

Solvent Extraction

Maceration

This involves soaking the plant material in a suitable solvent, filtering and concentrating the extract. The advantage of this method is that it uses cold solvent, which reduces decomposition, but it takes longer and uses greater volumes of solvent.

Percolation

This is a similar process but hot solvent is refluxed through the plant material. It is quicker and uses less solvent, but decomposition due to heat may occur. *Soxhlet extraction* is a form of continuous percolation with fresh solvent, and uses special glassware. The plant material is separated from the extract by encasing in a paper 'thimble' beneath dripping condensed solvent. When full, the solvent in the thimble siphons off into the main vessel containing the extract and the process continues. The advantage is that fresh solvent continually extracts the marc more efficiently with a minimum of solvent; however, heating is again a disadvantage.

Steam distillation

There is a special apparatus for distilling volatile oils which are immiscible with water. If the compounds being extracted are water-soluble, the method is less useful because a large volume of aqueous extract is produced; however, in some cases a partition system may be used to concentrate the extract.

Choice of Solvent

If the type of compounds being isolated is known, selective solvent extraction will make the process safer. If not, the usual way is to start with a non-polar solvent and exhaustively extract the marc, followed by a series of more polar solvents, until several extracts are obtained of increasing solute polarity. These may then be tested for biological activity. It is usual to remove fixed oils and fats using light petroleum, and to try not to extract starches (by avoiding aqueous solvents) in seed material, because these make further separation difficult and messy (*see* Figure 3.1).

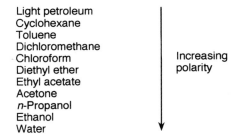

Light petroleum
Cyclohexane
Toluene
Dichloromethane
Chloroform
Diethyl ether
Ethyl acetate
Acetone
n-Propanol
Ethanol
Water

Increasing
polarity

Figure 3.1 List of solvents of increasing polarity, which also corresponds to their eluting power in chromatography.
Note that the extracts so obtained will need to be concentrated using a rotary vacuum evaporator. Generally, the more polar the solvent, the more heat is required to evaporate it, so the most volatile solvent that will be effective is chosen. Mixtures are often used, e.g. chloroform/methanol. Aqueous solvents will extract inorganic ions such as calcium, which may interfere with biological testing and will need removing by partitioning at some stage.

Further Purification

Once a biologically active extract is obtained, further purification may be undertaken and steps taken to identify the constituents. The majority of natural products used medicinally are of certain types and can be isolated and chemically tested as summarised below.

Alkaloids

These are organic bases containing nitrogen in a heterocyclic ring. Many have pronounced pharmacological activity. Examples of plants containing alkaloids are opium, ergot and rauwolfia. Further purification of alkaloidal extracts can be achieved by acid–base partitioning between aqueous and organic solvents, which exploits the different solubilities of the free base and salt form (Figure 3.2).

Several extraction methods are possible:

- *Aqueous acid.* 2% Sulphuric acid extracts the alkaloids as their salts. The acid extract is made alkaline with ammonia solution, and shaken gently with an organic solvent in a separating funnel. The alkaloids, as the free bases, partition in favour of the organic layer, leaving behind the unwanted non-basic substances.
- *70% Ethanol.* After extraction the solution can be concentrated to low volume, removing most of the ethanol in the process, acidified with dilute sulphuric acid and partitioned first with organic solvent to remove

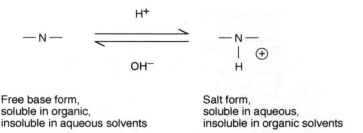

Figure 3.2   Acid–base partitioning: different solubilities of free base and salt forms.

unwanted acidic substances, then basified and re-extracted as above. This involves an extra purifying step and smaller volumes of solvents.

• *Alkali treatment.*   The plant material is moistened with aqueous ammonia or calcium hydroxide and extracted with organic solvent, with further partitioning as required.

These extracts can be used for chemical testing but if chloroform is used it must be completely removed before biological testing, as it has biological activity and disrupts cell membranes. Organic solvents can be 'back-washed' with purified water to remove traces of inorganic ions or other water-soluble material.

Quaternary alkaloids such as *N*-oxides which are always present in an ionised form are soluble in both aqueous and organic solvents, and cannot be partitioned as described. They may be precipitated with picric acid or Mayer's reagent and converted into soluble bromides or chloride using ion-exchange chromatography.

General Chemical Tests for Alkaloids

Add 1 drop of Mayer's reagent to the acidic extract: a pale precipitate indicates the presence of alkaloids. The same can be done with Dragendorff's reagent: an orange-red precipitate is positive. A drop of organic extract can be put onto a filter paper, allowed to dry and then sprayed with Dragendorff's, giving an orange spot if alkaloids are present. Tests for specific alkaloids can be found in *Clarke's Isolation and Identification of Drugs* (Moffat *et al.*, 1986).

Glycosides

These are a heterogenous group, since any compound linked to a sugar moiety is termed a glycoside. Their solubility, and hence extraction method, depends on the nature of the aglycone and the number and type of sugar molecules linked to it. Aglycones tend to be soluble in organic solvents, and the sugar part in aqueous solvents. Initial extraction is usually with ethanol or aqueous ethanol, but this dissolves many unwanted constituents and further purification, usually by partitioning, will be required.

Examples of pharmacologically active glycosides range from the simple phenolic compounds, e.g. salicin; to flavonoids, e.g. rutin; antraquinones, e.g. the sennosides; cardiac glycosides, e.g. digoxin; and saponins, e.g. the ginsenosides.

Most glycosides consist of the aglycone linked to the sugar by a hemiacetal link, C—O—C, but some have a C—N—C or C—S—C bond. These are readily hydrolysed by acid, but some are covalently bonded through a C—C bond. Hydrolysis in this case is with an oxidising agent such as ferric chloride added to the hydrolysing acid.

Chemical Tests for Glycosides

There is no general test apart from the presence of a sugar after hydrolysis. This can be assessed by making an aqueous extract, dividing into two parts, hydrolysing one portion and comparing these for the presence of reducing sugars, using Fehling's or some other test. It is not specific, since polysaccharides and other substances will give positive results, and is rarely used.

Saponins

Saponins are a particular form of triterpenoid glycosides and include the ginsenosides. They are so-called because of their soap-like effect, which is due to their surfactant properties, and this can be used as a test: if an aqueous extract of the plant material is shaken, a froth or foam will form if saponins are present. They also haemolyse red blood cells. To show this, an isotonic extract is made (using 0.9% NaCl, otherwise haemolysis will occur anyway because of osmotic pressure) and added to the blood. After centrifugation, any red colour in the supernatant means haemolysis has occurred.

Other Compounds

There are many other types of plant constituent, some of which fall into more than one category and which do not have specific tests or extraction methods. (For information on these, refer to 'Further Reading', particularly *Phytochemical Dictionary* and the pharmacognosy books.) Some examples are:

- *Terpenoids.* These are hydrocarbons, some of which are oxygenated or otherwise substituted. These include substances such as the saponins and cardiac glycosides which have already been mentioned, the iridoids, phytosterols, carotenoids, mono- and sesquiterpenes which are constituents of volatile oils, and others. They are more soluble in organic solvents and oils.

- *Phenolics.* These tend to be soluble in more polar solvent, such as water, alcohol or chloroform. This is a diverse group and includes simple phenolics, (like salicin which is also a glycoside), tannins, lignans, coumarins, flavonoids, benzofurans and anthocyanins.
- *Proteins and peptides.* These tend to denature easily and are more difficult to work with than other nitrogen-containing substances like alkaloids. They need to be extracted with cold water, separated with a gentle technique such as gel filtration or electrophoresis, and treated delicately at all times (i.e. not shaken, heated or acidified—think of egg white). Protein chemistry needs expertise and care.
- *Carbohydrates.* Once thought of as just nutrients rather than pharmacologically active substances, the mono- and disaccharides, polysaccharides and cyclitols are now considered to have useful properties, acting on the immune system as well as other body functions. Again, these are easily broken down, mainly by hydrolysis and are extracted gently in a similar way to proteins.

Separation of Compounds from an Extract

Once a biologically active extract has been obtained from the plant, it is necessary to fractionate the extract to find out where the biological activity lies. The more precise the isolation procedure the better, because it is more accurate to investigate the biological activity of a single compound than a mixed range, which may contain both agonists and antagonists in the same extract.

The usual method is some form of chromatography, and there are many sophisticated pieces of equipment with which to do this. However, many excellent results are still obtained using the old favourites, such as a silica gel or sephadex column, followed by thin-layer (TLC) or paper chromatography (PC). Chromatography was developed by natural product chemists working on plant material and is now used routinely in all areas of chemical, biological and medical research.

Chromatography

Chromatography is often defined as the separation of solutes in a mixture according to their affinity for a stationary phase, through which a mobile phase moves. It is used for analysis of small samples as well as on a larger scale for preparation of fractions. The mechanism by which separation occurs may be adsorption, partition, molecular diffusion or electrophoresis, or a combination of these. For producing fractions for initial pharmacological testing the most useful methods are column (CC), preparative thin layer (TLC), high performance liquid chromatography (HPLC) and gel filtration. The first three methods effect separation according to chemical affinity for

ionary and mobile phases. This may be either by adsorption, or parti-
here there is an inert support holding a stationary liquid phase over
an immiscible mobile phase moves. In HPLC there may be molecular
on involved to some extent. Gel filtration, as the name suggests,
ves molecular sieving, and separation depends on molecular size of
e, properties of the gel and the viscosity of the mobile phase.

tionary Phases

These are the same whether packed into a column or spread onto a plate,
but for large-scale work a column is the best starting point as the loading is
much greater. The extract is placed in as small a volume as possible on top
of the adsorbent and eluted through the column with progressively more
polar solvents. Adsorbents are also classified according to their polarity, the
more polar being the stronger adsorbents, and occasionally the weaker ones
are used as an inert support carrying a viscous liquid for a partition system.
The most common are given here:

Strong adsorbents: anhydrous alumina; silicic acid.
Medium polar adsorbents: silica gel; hydrous alumina; calcium carbonate.
Weak adsorbents: cellulose; keiselguhr.

Strong adsorbents are more likely to cause decomposition of solutes than
weak ones. In practice silica gel is the most commonly used of all, while for
highly labile compounds such as proteins, gel filtration will be necessary.
Solutes which are easily oxidised may be separated on a partition plate, for
example the phorbol esters are isolated using a diethylene glycol-impregnated
keiselguhr plate. In this case the plate is spread with the inert support and
developed in a tank containing the digol, before the extract is applied and run
in the chosen solvent system (which must not dissolve the stationary liquid
phase, of course). The same principle applies in reversed-phase HPLC, but the
stationary liquid phase, often a C_{18} hydrocarbon, is chemically bonded to the
support and again partition is the main mechanism of separation.

Selection of Fractions

From a column   With a preparative column fraction selection is arbitrary,
since visualisation is difficult and many fractions are obtained. Samples of
these fractions may be tested by analytical TLC and sprayed with a visualis-
ing reagent, so that similar ones can be pooled together. Eluates from gel
columns are treated similarly. The choice of TLC system and spray reagent
will depend on whether the constituents are known. Several 'universal'
sprays are used, and if necessary concentrated sulphuric acid with heat will
show up most organic compounds as a dark spot due to charring. There is a

comprehensive list of systems, solvents, reagents and a screening procedure given in *Plant Drug Analysis* (Wagner et al., 1983) *see* 'Further Reading').

From an HPLC Column HPLC, although a column method, uses a UV detector to monitor effluent from the column, so fractions can be selected as they come off the column. It is therefore most useful when the solutes have pronounced UV absorption, and cannot be used with solvents such as acetone which absorb UV. HPLC is not suitable for crude extracts or particulate matter which will damage the column, and is better used for further fractionation of filtered, fairly clean mixtures. It is important to remember to redistil and de-gas solvents also.

From TLC Preparative ('prep') TLC is used in the same way as analytical TLC, except the layer is thicker and takes a higher loading of extract. The extract is spotted continuously along the base-line and the plate developed in the usual way. Resolution with 'home-made' plates is nowhere near as good as with commercially available ones and this must be considered when transferring from an analytical to a prep scale. Visualisation is by spraying a narrow strip at each side of the plate, while masking the main part. The zones containing the required fractions are scraped off, taking care not to include any of the sprayed part, and the solutes eluted with a suitable solvent. These fractions can be concentrated down and used for biological testing, combining the same zone from as many plates as possible.

The other, more sophisticated, techniques like gel electrophoresis are not normally used for crude plant extracts so will not be covered here.

Solubilisation for Biological Testing

All solvents should be tested alone in the system and, assuming no pharmacological activity, then used as a control during the test. The extraction procedure chosen will have given a good indication of the solubility of the extract. Solvents chosen must have the following properties: lack of biological activity; high solubility for plant extract; miscibility with physiological solution used in the test. Suitable ones include ethanol, water, acetone, dimethyl sulphoxide and mixtures of these. A lipophilic extract can often be solubilised with a detergent such as Tween 80.

Summary of a Typical Procedure

1. Make crude extract(s), non-polar, polar etc.; put through biological test system. Select active extract(s).
2. Fractionate crude extract using column chromatography, to give fractions F1, F2, F3, etc. Test with TLC and pool those which are chemically similar. Put through biological test and select active fractions.

3. Further purify by prep TLC or HPLC. Isolate single spots or peaks and test biologically.
4. Use analytical TLC, on at least two different systems, to see if spot or peak consists of more than one compound. If so, further prep TLC should be carried out to get a single compound.
5. Use for structure elucidation and measure biological activity quantitatively, e.g. ED_{50}. Use for further investigations into mechanism of action, structure–activity relationships and so on.

FURTHER READING

Pharmacognosy

Evans, W.C. (1989) *Trease and Evans' Textbook of Pharmacognosy*, 13th Edn. Baillière Tindall, London.
Steinegger, E. and Hansel, R. (1988) *Lehrbuch der Pharmacognosie und Phytopharmazie*, 4th Edn. Springer-Verlag, Berlin.
Tyler, V.E., Brady, L.R. and Robbers, J.E. (1988) *Pharmacognosy*, 9th Edn. Lea and Febiger, Philadelphia.
Williamson, E.M. and Evans, F.J. (1988) *Potter's New Cyclopaedia of Botanical Drugs and Preparations*. C.W. Daniels, Saffron Walden.

To Check Structures Against Known Compounds

Budavari, S., Ed (1989) *The Merck Index: An Encyclopedia of Chemicals, Drugs and Biologicals*, 11th Edn. Merck and Co. Inc, Rahway, NJ.
Harborne, J.B. and Baxter, H., Eds (1993) *Phytochemical Dictionary: A Handbook of Bioactive Compounds from Plants*. Taylor and Francis, London.

Analysis of Plant Material

Hostettmann, K., Hostettmann, M. and Marston, A. (1986) *Preparative Chromatography Techniques: Applications in Natural Product Isolation*. Springer-Verlag, Heidelberg.
Moffat, A.C., Jackson, J.V., Moss, M.S. and Widdop, B., Eds (1986) *Clarke's Identification of Drugs*, 2nd Edn. The Pharmaceutical Press, London.
Wagner, H., Bladt, S. and Zgainski, E.M., Translated by Scott, A. (1983) *Plant Drug Analysis*. Springer-Verlag, Heidelberg.
Waterman, P.G. and Mole, S. (1994) *Analysis of Phenolic Plant Metabolites*. Blackwell Scientific Publications, Oxford.
Williams, D.H. and Fleming, I. (1989) *Spectroscopic Methods in Organic Chemistry*, 4th Edn. McGraw-Hill, London.

4

The Gastro-intestinal Tract

Medicines originating from plants form a very important part of the treatment of gastro-intestinal diseases. The main disorders encountered are dyspepsia (indigestion); spasms of the intestine such as colic; peptic and duodenal ulceration; nausea and vomiting; constipation and diarrhoea.

Animal models exist with which to test plant extracts for activity against some of these conditions. These mainly involve testing the extract for antispasmodic or carminative activity as predictions for potential usefulness in the treatment of dyspepsia and colic; and the ability to protect against ulcer formation induced by alcohol, aspirin and other ulcerogenic drugs is indicative of anti-ulcer activity. Antinauseant, anti-emetic, and antidiarrhoeal activities are more difficult to test for, since animals show different sensitivities to substances causing these conditions, and nausea is a subjective disorder. However, experimental techniques do exist and will be described where possible.

Screening programmes testing for laxative activity are not seen nowadays and are not really needed any longer, since the plant kingdom already provides most of our most widely used laxatives; the stimulant laxatives senna, frangula, cascara and aloes, and the bulk/fibre laxatives isphagula, psyllium and bran are examples. A simple test for laxative activity, if required, is given under the *in vivo* methods in the antispasmodics section, this Chapter.

TESTING FOR ANTISPASMODIC ACTIVITY

Antispasmodics, often called spasmolytics or carminatives, are used to ease griping or colicky pains, expel wind (flatus) and relieve dyspepsia or indigestion. The plant kingdom is rich in such compounds; in fact most remedies used in conventional medicine include at least one antispasmodic of plant origin, even if only the ubiquitous peppermint oil!

Preparations used as antispasmodics in the gastro-intestinal tract may also be used for other disorders (e.g. papaverine is also used as an antitussive) since they are utilising a common smooth muscle relaxant effect. Hyoscine, an important antispasmodic, is used also as an anti-emetic and to treat dysmenorrhoea, and exerts its effects via its anticholinergic activity. Drugs such as these will therefore have effects on bodily systems other than the gastro-intestinal tract; these will be classed as side-effects and must be taken into account.

Obviously, antispasmodics are used to treat colic and indigestion, and those which have a powerful inhibiting effect on peristalsis are particularly useful for treating diarrhoea, such as the opium alkaloids morphine and codeine.

Examples of plant-derived antispasmodics include: some tropane alkaloids (atropine, hyoscine (scopolamine), hyoscyamine); opium alkaloids (papaverine, codeine); essential oils (peppermint, dill, caraway, anise, thyme, garlic, chamomile); and flavonoids (apigenin, kaempferol, quercetin).

Animal Models: Antispasmodics

In vitro testing is usually carried out by observing any direct effect on isolated intestine preparations in an organ bath, e.g. guinea-pig ileum, rat duodenum or rabbit jejunum. Spasms are induced by a number of agonists, the nature of which may give a good indication of the mechanism of action. There are some *in vivo* methods, e.g. propulsion of a charcoal meal in the mouse, which will be described briefly; inhibition of caster oil-induced diarrhoea; and intestinal movement in the cat. These obviously require compliance with animal experimentation legislation since they are dealing with live animals.

A non-animal, physical method for testing carminative action has been described (Harries *et al*, 1978; see 'Further Reading') involving measuring anti-foaming activity *in vitro*. Although not related to antispasmodic activity, it was found that a reduction in gastro-intestinal foam is a property of many carminatives, and of course public opinion is in favour of any reduction in animal usage.

Materials and Methods: Antispasmodics (Figures 4.1–4.5; Tables 4.1, 4.2)

IN VITRO METHODS

Isolated Guinea-pig Ileum and Related Preparations

This is a robust, simple preparation that can be used to obtain much useful information. It is as well to ensure that it is properly set up (*see* section on peripheral nervous system, Chapter 10, for further details).

Apparatus

A segment of ileum 2–4 cm long from a freshly killed guinea-pig is suspended in an organ bath (5–15 ml) containing Tyrode solution (pH 7.4) at 34–37°C, aerated with 5% CO_2, 95% O_2, loaded with a suitable weight or tension (about 0.5 g). The longitudinal contractions are recorded on a kymograph or Dynograph recorder running at a speed of about 5 mm/min.

Reagents

Tyrode solution: *see* formula in Appendix I.

Agonists (Spasmogens)

Molar concentrations may need to be varied:
Acetylcholine (ACh) $1 \times 10^{-8} - 1 \times 10^{-6}$ M.
Barium chloride ($BaCl_2$) $1 \times 10^{-5} - 2.5 \times 10^{-3}$ M.
Histamine $1 \times 10^{-8} - 1 \times 10^{-5}$ M.
Prostaglandin E_2 (PGE_2) approx. 30 nM.
Others: e.g. leukotriene D_4, calcium chloride.
For use of 5-hydroxytryptamine (5-HT, serotonin), *see* anti-emetic section, this Chapter.

Antagonists (Antispasmodics, Spasmolytics)

For comparison: papaverine $1 \times 10^{-5} - 1 \times 10^{-4}$ M.
Test plant extract: dissolved in dimethylsulphoxide (DMSO), alcohol, Tyrode's solution; may be emulsified with Tween 80 if necessary.

Method

Set up apparatus as described and allow to equilibrate for about 1 h. Produce several control contractions with the agonist(s) of choice, and if required a dose of papaverine, 2–5 min prior to agonist to demonstrate antispasmodic activity. Check that solvents and plant extracts have no unexpected activity. Antispasmodics under test may be used to pretest the ileum segment or to attempt to reverse activity. At least six experiments are needed to give a valid result; do not forget to wash out well after each experiment. The response is expressed as a percentage of the contraction obtained before addition of the antagonist; therefore it may produce a relaxation giving a result higher than 100%. A dose representing a 50% response (ED_{50}) of the antispasmodic can be obtained from the graph, and is usually referred to as the IC_{50} (inhibitory concentration 50%). Examples of the tracing obtained and the dose–response curve which may be drawn are shown below. Isolated rat duodenum can be used in a similar way; and isolated rabbit jejunum (3–4 cm long), which is spontaneously contracting, is useful as agonists are not needed. Change in amplitude of contractions may be used as a measure of response and plotted in the usual way.

IN VIVO METHODS

Gastro-intestinal Transit of a Charcoal Meal

This is one of the few *in vivo* methods available. It involves measuring the transit time of a charcoal meal, which is easily visible, through the gastro-intestinal tract of mice. The method given would need modifying if other animals were to be used.

The animals are fasted for 24 h prior to the experiment, pretreated with the test plant extract and 15 min later given orally the charcoal meal, which consists of 0.4 ml of an aqueous suspension of charcoal in 5% gum acacia. The animals are killed by ether inhalation 20 min after the meal, and the intestines and stomach removed. The pylorus is attached to a glass rod and the intestine suspended for 20 sec with a weight of 3 g attached to the ileo-caecal junction, to straighten it out. The mean distance travelled by the charcoal can then be measured and compared with the control group, expressing results as a percentage of the total length of intestine. In the method given, the test plant extract was garlic oil (0.1 ml/kg), which produced an approximate reduction of 75% in the gastro-intestinal propulsion of the charcoal meal. An increase in rate of propulsion would indicate a laxative effect.

Castor Oil-induced Diarrhoea

Administration of castor oil induces diarrhoea in mice (90% of animals) within 4 h. Pretreatment with an antidiarrhoeal plant extract will reduce the number of animals exhibiting diarrhoea and also reduce such incidents.

Mice are divided into two groups of 30 animals each; the first (control) group receiving a suitable placebo, in a similar dose and volume, to the test group receiving plant extract. 30 Min later both groups are treated with castor oil (0.5 ml/animal, p.o.) and each mouse kept for observation under a glass funnel on tissue paper. Onset of diarrhoea and the number of such episodes are noted for each animal, for a total of 6 h. Statistical comparison of the mean values of total diarrhoeal episodes in control and test groups are made using Student's *t*-test.

Results can be expressed graphically, as shown below. For full details of both experiments, see Joshi, D. J. *et al* (1987), *Phytotherapy Research* 1 (3) 140–141.

Examples of Traces

With Isolated Guinea-pig Ileum

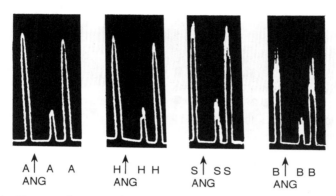

Figure 4.1 Non-specific spasmolytic effect of an antagonist, angelicin (ANG 15 μg/ ml) against the agonists acetylcholine (A), histamine (H), serotonin (S) and BaCl$_2$ (B) induced contractions in isolated guinea-pig ileum preparation. From Spasmolytic activity of angelicin: a coumarin from *Heracleum thomsoni*. Patnaik, G.K. *et al* (1981), *Planta Med*, **53**, 517–52, with permission.

With Isolated Rabbit Jejunum

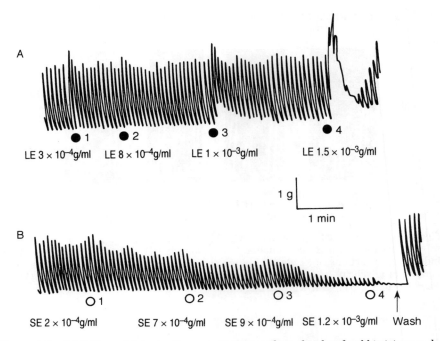

Figure 4.2 Inhibition of tone and spontaneous motility of isolated rabbit jejunum by different concentrations of angelicin (0.25 µg/ml, first panel; 1 µg/ml, second panel). From Spasmolytic activity of angelicin: a coumarin from *Heracleum thomsoni*. Patnaik, G.K. *et al* (1981), *Planta Medica*, **53**, 517–552, with permission.

Treatment of Results

Dose–Response Curve

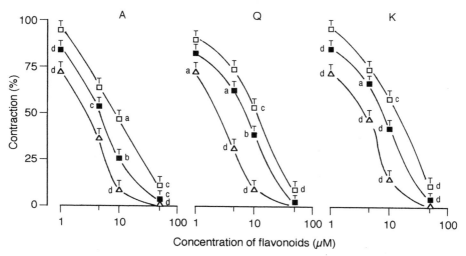

Figure 4.3 Concentration- and time-dependent effect of apigenin (A), quercetin (Q) and kaempferol (K) on (PGE$_2$) prostaglandin-induced contraction of guinea-pig ileum. Responses are percentages of control contraction to 30 nM PGE$_2$. Ileum was incubated with each flavonoid for 2 (□), 5 (■) or 10 (△) min before adding agonist. Each point represents the mean $\pm$ SEM of five experiments. For statistical significance: [a]$p < 0.05$, [b]$p < 0.01$, [c]$p < 0.005$, [d]$p < 0.001$. From Reduction of agonist-induced contractions of guinea-pig isolated ileum by flavonoids. Capasso, A. *et al* (1991), *Phytother. Res.*, **5**(2), 85–87, with permission.

Alternative Expression of Results: Dose–Response Represented as a Bar-chart

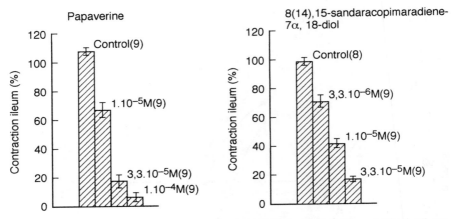

Figure 4.4   Effect of increasing concentrations of papaverine (left) and 8(14),15-sandaracopimaradiene-7α,18-diol (right) on the contraction induced by barium chloride $(2.3 \times 10^{-3} \, M)$ in the guinea-pig ileum. The contractions are expressed as a percentage of the maximal contraction obtained in the same tissue before the administration of the antispasmodic. The bracketed figure indicates the number of experiments. From Active principles of *Tetradenia riparia*; II. Antispasmodic activity of 8(14),15-sandaracopimaradiene-7α, 18-diol. Van Puyvelde, L. *et al* (1987), *Planta Medica*, **53**, 156–158, with permission.

Comparison of Anti-spasmodic Action of Different Drugs as a Table

Table 4.1   Spasmolytic activity (expressed as IC_{50} and E_{max}) of *Thymus longiflorus* Boiss essential oil and its major components (mean $\pm$ SEM of the results in five animals per assay)

Drugs	IC_{50} (μg/ml) $\pm$ SEM	$E_{max} \pm$ SEM
Essential oil of *T. longiflorus*	94.14 $\pm$ 4.69	100
α-pinene	3.89 $\pm$ 0.29	86.63 $\pm$ 2.88
Camphene	5.08 $\pm$ 1.29	91.47 $\pm$ 1.88
Myrcene + α-phellandrene	9.67 $\pm$ 1.79	90.06 $\pm$ 2.79
Terpenic fraction[a]	5.05 $\pm$ 0.48	100
Cineole	34.79 $\pm$ 2.95	100

[a] Terpenic fraction: mixture of the terpenes in the same relative proportions as those contained in the essential oil. From Spasmolytic action of the essential oil of *Thymus longiflorus* Boiss in rats. Zarzuelo, A. *et al* (1989), *Phytother. Res.*, 3(1), 36–37, with permission.

Testing for Antispasmodic Activity in the Gastro-intestinal Tract in vivo: *Inhibition of Castor Oil-induced Diarrhoea*

Table 4.2 Effect of garlic oil on the severity of castor oil-induced diarrhoea in mice (mean $\pm$ SEM; $n = 30$)

Group	Treatment	No. of diarrhoeal episodes						Total no. of diarrhoeal episodes
		1 h	2 h	3 h	4 h	5 h	6 h	
1	Groundnut oil (control) (0.1 ml/10 g, p.o.)	0.60 $\pm$ 0.17	1.10 $\pm$ 0.19	1.18 $\pm$ 0.13	1.25 $\pm$ 0.11	1.35 $\pm$ 0.10	1.21 0.10	6.69 $\pm$ 0.37
2	Garlic oil (0.1 ml/10 g, p.o.)	0.00[a]	0.00[a]	0.00[a]	0.06 $\pm$ 0.04[a]	0.17 $\pm$ 0.07[a]	0.17 $\pm$ 0.07[a]	0.40 $\pm$ 0.11[a]

[a] $p < 0.001$, in comparison with the control group. Student's 't' test. From Gastrointestinal actions of garlic oil. Joshi, D.J. *et al* (1987) *Phytother. Res.*, 1(3), 140–141, with permission.

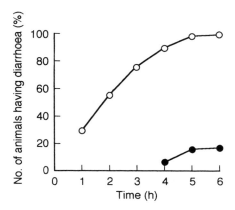

Figure 4.5 Animals having diarrhoea at various time intervals after administration of castor oil in control (○—○) and garlic oil treated (●—●) mice. $n = 30$ in each group. From Gastrointestinal actions of garlic oil. Joshi, D.J. *et al* (1987) *Phytother. Res.*, 1(3), 140–141, with permission.

ANTI-EMETIC TESTING

The search for new anti-emetics continues, not only for the treatment of the usual types of sickness, including travel sickness, but especially for the nausea and vomiting induced by cancer chemotherapy. Some cytotoxic drugs cause emesis of such severity that it affects the acceptability of such treatment to the patient. Natural products are already used as anti-emetics, e.g. hyoscine for travel sickness, dried ginger rhizome for travel sickness and vomiting of pregnancy, and the semi-synthetic cannabinoid derivative nabilone for cytotoxic-induced vomiting.

Animal Models: Anti-emetics

Testing for activity against nausea and vomiting can be done in a number of ways, e.g. using the antispasmodic tests already described. Anticholinergics are commonly used to prevent travel sickness. It is useful to test for anti-emetics in this way using the guinea-pig ileum isolated organ preparation, with the agonist serotonin (5-hydroxytryptamine, 5-HT), since there is a correlation between 5-HT$_3$ receptor antagonist potency and anti-emetic efficacy in the same way as dopamine D$_2$ receptor blocking activity reflects anti-emetic activity (*see* Ison, 1986; Fozord, 1987; in 'Further Reading'). There is also an *in vivo* animal model described, using emesis in the frog.

Conventional ways of testing utilise the compulsive gnawing response in rats and the inhibition of the pecking response in pigeons, both of which are induced by apomorphine and the abolition indicative of anti-emetic activity (*see* 'Further Reading').

Materials and Methods: Anti-emetics

IN VITRO METHODS

Antispasmodic and Antiserotonergic Testing (Table 4.3)

Apparatus and Reagents

Any of the previous antispasmodic preparations may be used. To test specifically for antiserotonergic activity, the isolated guinea-pig ileum preparation described uses a modified Krebs Ringer solution in the organ bath; (mM): NaCl 120.4; KCl 4.8; MgSO$_4$ 1.3; CaCl$_2$ 1.2; KH$_2$SO$_4$ 1.2; NaHCO$_3$ 25.2; and glucose 5.8. This must be gassed with a mixture of 95% O$_2$, 5% CO$_2$.

Serotonin (5-HT) is the agonist.

Antagonists

5-HT$_3$ receptor antagonist ondansetron.
5-HT$_3$ and dopamine D$_2$ receptor antagonist: metoclopramide.
Acetone extract of ginger (*Zingiber officinale*), 50 µg/ml, can be used as an example of the kind of result to expect.

Method

As described previously for isolated guinea-pig ileum. Serotonin (5-HT) is used as the agonist, added to the bath at concentrations of 10^{-5} M to obtain the contractile response, and the tissue washed out three times every 10 min for the next 30 min. The test drug is then added, and 20 min later a suitable dose of 5-HT is added, and the response compared to that obtained by the first contraction of 5-HT, taken as 100%. Then use other agonists in addition, to determine whether the effect of the plant extract is specific or non-specific.

For treatment of results *see* antispasmodic section, this Chapter, and Table 4.3.

IN VIVO METHOD

Emesis in the Frog

The method given is that of Maki *et al* (1987) (see below). Frogs (normal or ranid) are divided into groups of 5 frogs each, and fed small earthworms on the day before the experiment. The test extract, dissolved in 0.2 ml, is administered orally and the frogs allowed to stabilise for 30 min. Then the agonist, apomorphine (1 mg per 0.2 ml), is administered orally, and the behaviour of the frogs observed for 1 h at intervals of 5 min to see whether gastric contents are vomited out inside their mouths (the frogs may not actually expel the stomach contents but keep them inside the mouth with the tongue, necessitating a compulsory opening of the mouth to view). The percentage inhibition is calculated from the total number of frogs tested and those showing no signs of vomiting. For further details *see* Maki T *et al* (1987), *Planta Med.*, **53**, 410–414; Maki T *et al* (1985), *J. Agric. Food. Chem.*, **33**, 1204.

Treatment of Results

Table 4.3 Effects of *Zingiberis rhizoma* (acetone extract), active fractions and gingerols on the contractile action of 5-HT $(10^{-5}\,\text{M})$ in the guinea-pig ileum

Compounds	Concentration	Relative tension (% of control $\pm$ SEM)
Control	0.1% DMSO	100.0 ± 9.6
Acetone extract	5 µg/ml	75.5 ± 8.6
	25 µg/ml	18.0 ± 0.1^{a}
	50 µg/ml	5.7 ± 1.1^{b}
Fraction 2	16.3 µg/ml	12.0 ± 3.6^{b}
Fraction 3	26.2 µg/ml	17.2 ± 7.5^{b}
[6]-Gingerol	$3 \times 10^{-5}\,\text{M}$	59.1 ± 7.2^{a}
[8]-Gingerol	$10^{-6}\,\text{M}$	87.9 ± 6.1
	$10^{-5}\,\text{M}$	13.6 ± 5.4^{b}
	$3 \times 10^{-5}\,\text{M}$	0.0 ± 0.0^{b}
[10]-Gingerol	$3 \times 10^{-5}\,\text{M}$	13.7 ± 6.0^{b}
Cocaine	$10^{-5}\,\text{M}$	62.6 ± 6.3^{b}
	$3 \times 10^{-5}\,\text{M}$	44.4 ± 9.3^{b}

Control was effects of 25 µl-DMSO in the 25 ml-tissue bath. Each value represents the mean $\pm$ SE of 5–8 experiments ($^{a}p < 0.05$, $^{b}p < 0.01$). From Active components of ginger exhibiting anti-serotonergic actions. Yamahara, J. (1987), *Phytother. Res.*, 3(2), 70–71, with permission.

TESTING FOR ANTI-ULCER ACTIVITY

Drugs used to treat or prevent ulcers act in a variety of ways, either pharmacologically, by reducing acid secretion, or by physically forming a protective barrier for the digestive mucosa. There is also involvement of *Helicobacter pylori*, although the presence of this organism does not automatically lead to ulceration. Anticholinergics (e.g. atropine, hyoscine and the newer, more specific synthetic anticholinergic drugs) reduce intragastric acidity and reduce spasm; histamine H_2-receptor antagonists reduce acid secretion by the parietal cells and proton pump inhibitors block acid secretion.

Gastric and duodenal ulcers are common conditions, and occur in degrees of severity from simple gastritis (inflammation of the gastric mucosa) to manifest ulcers observable by X-rays.

The increase in acidity may be caused by stress (mental and physical); diet; alcohol; and treatments such as non-steroidal anti-inflammatory drugs. There are therefore several possible approaches to the treatment and prevention of ulcers, but we are concerned only with drug treatment here.

Obviously, to test the effectiveness of a drug it is necessary to induce ulcers experimentally, and this may be done by mimicking the ulcerogenic stimuli already mentioned.

Plant drugs with anti-ulcerogenic activity include the tropane alkaloids, atropine and hyoscine, which are anticholinergic; liquorice, the mode of action of which is not fully understood; flavonoid-containing drugs; and other less well known substances such as gefarnate (from white cabbage juice) and extract of unripe plantain.

Animal Models: Anti-ulcer Drugs

The most common animal models use aspirin (acetylsalicylic acid), alcohol and stress to induce ulceration. Drugs may be tested by pretreatment or concurrent administration, to assess preventative potential, or after ulceration to test curative ability. Other tests, such as measurement of gastric acid secretion or DNA levels, may be carried out to help elucidate the mechanism of action. The route of administration of the drug is important, e.g. to show whether there is a genuine pharmacological effect rather than just a dilution of the ulcerogen in the gastro-intestinal system, or whether the plant extract is acting as a physical protectant—obviously this would not be the case if not given orally.

Chemically Induced Ulcers

Ethanol

Alcohol is probably responsible for a significant proportion of human ulcers, and so is a suitable experimental ulcerogenic stimulus. Ethanol is thought to induce ulcers independently of any effect on gastric acid secretion, since prostaglandins, which are potent inhibitors of ethanol-induced ulceration, are cytoprotective at doses too small to modify gastric acid secretion. Histamine H_2-receptor antagonists are equally effective at inhibiting ethanol-induced lesions; they are not cytoprotective but act by preventing acid secretion. The prostaglandins, especially PGE_2, and their synthetic analogues, increase the amount of mucus secreted. Any involvement of endogenous prostaglandins can be tested by pretreatment with a prostaglandin synthetase inhibitor (such as indomethacin) in a subulcerogenic dose.

Aspirin (Acetylsalicylic Acid)

Aspirin and related non-steroidal anti-inflammatory drugs (NSAIDs) cause some of their ulcerogenic effects by inhibiting production of the cytoprotective prostaglandins. Unfortunately, since NSAIDs are of paramount importance in the treatment of rheumatoid conditions, it is essential to develop protective means to combat this, and aspirin-induced ulceration is therefore an important animal model. The full explanation regarding NSAID gastro-intestinal damage is not clear, though the inhibition of prostaglandin synthesis is very important. Some cytoprotective agents, including flavonoids, appear to work against this damage by preventing the associated inhibition of mucus production, but without stimulating cell proliferation. With NSAID-induced ulcers, H_2-receptor blockers will repair the damage; however, cyto-protectants such as flavonoids will also increase mucus secretion, as measured by an increase in sialic acid (*N*-acetylneuraminic acid, NANA, a component of mucoproteins). Certain flavonoids, e.g. hypolaetin-8-glucoside, have anti-inflammatory as well as anti-ulcer activity; therefore, it should not be assumed that such actions are mutually exclusive or that all forms of anti-inflammatory activity are similar in profile to that of the synthetic NSAIDs. At present, however, only certain natural products, e.g. some of the flavonoids and other phenolic glycosides, such as from *Filipendula ulmaria*, liquorice derivatives and nimbidin, appear to have anti-inflammatory actions coupled with a beneficial rather than a damaging effect on the gastric mucosa.

Other Chemically Induced Methods of Ulcer Formation

Histamine, indomethacin, serotonin (5-hydroxytryptamine, 5-HT) and corti-costeroids (e.g. prednisolone) will all induce gastro-intestinal ulcers when given in suitable dosages and conditions. H_2-receptor antagonists will prevent those caused by histamine, which in animal models tend to be duodenal ulcers, and indomethacin is an NSAID which works in a similar way to aspirin. The picture with serotonin is more complex, and since it is a brain transmitter involved in stress, trauma and depression, it is important in the aetiology of gastric ulcers. However, the pharmacological diversity of serotonin renders it out of the scope of this book.

Stress-induced Ulcer Formation

Producing stress-induced ulcers in animals is obviously unpleasant, but since stress is a major cause of ulcers in man it is necessary to occasionally use an animal model in which ulceration is induced by stress. In the example given, immobilisation and cold are used as the stimuli.

Materials and Methods: Anti-ulcer Drugs (Figure 4.6; Tables 4.4, 4.5)

The basic method applies to all ulcerogenic stimuli, chemical and physical (i.e. stress).

Animals

A group consists of at least six animals, usually rats (male, Wistar) but guinea-pigs are used occasionally, treated as follows:

Control groups
(a) Normal saline only, administered orally (p.o.)
(b) Ulcerogen in saline, usually administered p.o.
(c) Ulcerogenic stress procedure alone.

Test groups.   Plant extract in a suitable dose range (*see* below), administered by method of choice, such as orally, intraperitoneally (i.p.) or subcutaneously (s.c.), with ulcerogen.

Standard reference group, if required.   Administer a known anti-ulcer drug, e.g. liquorice or ranitidine, with ulcerogen.

Reagents

Ulcerogens
Ethanol 1 ml/kg body weight, p.o.
Aspirin 200 mg/kg p.o.
Standard anti-ulcerogens.
Liquorice extract 200 mg/kg p.o.
Gefarnate 50 mg/kg p.o.
Histamine H_2-receptor antagonists: cimetidine or ranitidine 100 mg/kg.
Test solutions.
Plant extract suspended in normal saline, dose range e.g. 50–500 mg/kg, administered p.o., i.p. or s.c. as required.
Plant extract should have been previously tested for toxicity (including gastro-intestinal damage).
Stress procedure
Immobilisation and cold: the rat is immobilised in a cylindrical cage and maintained at 3–5°C for 3 h.

Method

Pre-treat animals by administering test drug, standard reference drug if required, and control saline by chosen route, at least 0.5–1 h before ulcerogenic procedure.

Induce ulcers by administering ulcerogen (ethanol, aspirin, etc.) orally or carrying out stress procedure.

After 1 h, sacrifice animals and remove stomach (or duodenum in the case of histamine-induced ulcers), slit along curvature and open to assess ulcer damage.

Note: If it is intended to measure gastric secretion, then the stomach should be ligated before removal, and the contents drained and collected (including washing out with saline), before slitting along curvature to assess ulceration.

Assessment of Ulcer Damage

A scale of severity or 'ulcer index' needs to be determined to assess damage to the mucosa, in such a way that number and size of lesions is taken into account. Stress-induced ulceration tends to be less severe than that produced, for example, by aspirin. The scale chosen would depend on the extent and range of lesions produced; however, examples would be to grade ulcers as follows:

Grade/description
0 = No ulcer
1 = Haemorrhagic and slightly dispersed ulcers, less than 2 mm length
2 = One ulcer as above, up to 5 mm length
3 = More than one ulcer grade 2
4 = One ulcer above 5 mm in length
5 = More than one grade 4 ulcer

etc. to:

10 = Total ulceration and haemorrhage

Or a simpler scale:

0 = Absence of damage
1 = Redness of mucosa
2 = Erosion
3 = Ulcer

For statistical treatment of results, *see* examples of actual results below and references cited there.

Associated Experiments

Estimation of sialic acid (NANA) content can be carried out to assess mucus formation as an indication of cytoprotective activity and mechanism of anti-ulcer action. Total acidity can also be determined as a measurement of gastric secretion. DNA determination gives an indication of stimulation of cell proliferation. For methods, *see* 'Further Reading'.

Treatment of Results

Ethanol-induced Gastric Lesions

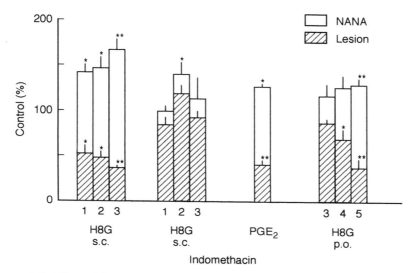

Figure 4.6 Effect of drug treatments on the area of lesion and total NANA content in the ethanol-induced lesions test. H-8-G: $1 = 60$, $2 = 80$, $3 = 100$, $4 = 200$, $5 = 300$ mg/kg. PGE_2: 100 µg/kg s.c. From Anti-ulcer activity of hypolaetin-8-glucoside (H-8-G). Alcarez, M.J. and Tordera, M. (1988), *Phytother. Res.*, 2(2), 85–88, with permission.

Figure 4.6 shows the influence of drug treatments on gastric lesion and sialic acid (NANA) content as a measure of mucus formation. The administration of H-8-G (s.c. or p.o.) to rats prevented the formation of lesions in a dose-related manner. The oral doses needed for protection were higher. Thus, the s.c. dose required to reduce mucosal damage to 50% of the control value (ED_{50}) was 67.92 mg/kg, whereas ED_{50} by oral administration was 266.68 mg/kg.

Gastric Ulcer Induced by Other Agents

Table 4.4 Acute gastric ulcer induced by absolute EtOH

Groups	Number of stomachs/ score	Incidence of ulcer (%)	Index ulcer (mm^2)
Control group ($n = 9$)	1/3 1/4 1/5 4/6 2/8	100	70.83 ± 19.36
Gefarnate 50 mg/kg ($n = 10$)	1/1 1/2 1/5 5/6 2/8	100	51.88 ± 10.05[b]
Ether extract 500 mg/kg ($n = 10$)	4/1 1/2 1/3 1/5 3/6	100	33.03 ± 10.77[a]
Quercetin 100 mg/kg ($n = 10$)	4/1 1/2 3/6 2/8	100	59.50 ± 18.11[b]

Mann–Whitney U-test: [a] $p < 0.05$, [b] not significant. From Anti-ulcerogenicity of the flavonoid fraction from **Dittrichia viscosa** (L.) W. Greuter in rats. Martin, M.J. *et al* (1988), *Phytother. Res.*, **2**(4), 183–186, with permission.

Table 4.5 Acute gastric ulcer induced by immobilisation and cold

Groups	Number of stomachs/ score	Incidence of ulcer (%)	Index ulcer (mm^2)
Control group ($n = 9$)	2/2 3/3 4/4	100	3.23 ± 0.56
Ranitidine 100 mg/kg ($n = 7$)	5/0 2/2	29	0.18 ± 0.05[a]
Ether extract 500 mg/kg ($n = 9$)	4/0 1/1 3/2 1/4	56	0.76 ± 0.24[b]
Aqueous extract 500 mg/kg ($n = 9$)	2/0 1/1 3/2 3/2	78	1.42 ± 0.41[b]
Ethyl acetate extract 500 mg/kg ($n = 10$)	2/0 3/2 3/3 2/4	80	1.50 ± 0.40[b]
Quercetin 100 mg/kg ($n = 10$)	3/0 1/1 2/2 2/3 2/4	70	1.41 ± 0.37[b]

Mann–Whitney U-test: [a] $p < 0.01$, [b] $p < 0.05$. From Anti-ulcerogenicity of the flavonoid fraction from **Dittrichia viscosa** (L.) W. Greuter in rats. Martin, M.J. *et al* (1988), *Phytother. Res.*, 2(4), 183–186, with permission.

FURTHER READING

See standard textbook list (Appendix I) and references in the text, and:

Antispasmodic Activity

Famaey, J.P. *et al* (1977). Effect of high concentrations of non-steroidal and steroidal anti-inflammatory agents on prostaglandin-induced contractions of the guinea-pig isolated ileum. *Prostaglandins*, 13, 107–114.

Fanning, M.J. *et al* (1983). Quercetin inhibits anaphylactic contraction of guinea-pig ileum smooth muscle. *Int. Arch. Allergy Appl. Immun.*, 71, 371–373.

Harries, N. *et al* (1978). Anti-foaming and carminative actions of volatile oils. *J. Clin. Pharm.*, 2, 171–177.

Johnson, L.R., Ed. (1987). Extrinsic control of digestive tract motility. In *Physiology of the gastrointestinal tract*, Vol. 1, Raven Press, New York, pp. 507–553.

Kromer, W. (1988). Endogenous and exogenous opioids in the control of gastrointestinal motility and secretion. *Pharmacol. Rev.*, **40**, 121–162.

Meli, R. *et al* (1990). Inhibitory action of quercetin on intestinal transit in mice. *Phytother. Res.*, **4**, 201–202.

Van Rossum, J.M. (1963). Cumulative dose–response curves II. Technique for the making of dose–response curves in isolated organs and the evaluation of drug parameters. *Arch. Int. Pharmadyn. Ther.*, **143**, 299–232.

Antidiarrhoeal Testing

Buchheit, K.H. (1989). Inhibition of cholera toxin-induced intestinal secretion by the 5-HT$_3$ receptor antagonist ICS 205-930. *Naunyn Schmiedebergs Arch. Pharmacol.*, **339**, 704–705.

Dettmar, P.W. *et al* (1989). α_2-Adrenoceptor regulation of electrolyte transport in rat jejunum. *Br. J. Pharmacol.*, **88**, 32.

Doherty, N.S. and Hancock, A.A. (1983). Role of α_2-adrenergic receptors in the control of diarrhoea and drug therapy. *J. Pharm. Exp. Ther.*, **225**, 269–274.

Feldman, S. and Pickering, L.K. (1981). The relationship of diarrhoea and drug therapy. *Trends Pharm. Sci.*, **2**, 165–167.

Ruwart, M.J. *et al* (1980). Clonidine delays small intestinal transit in the rat. *J. Pharm. Exp. Ther.*, **212**, 487–490.

Thomas, G. *et al* (1992). Further studies on the antidiarrhoeal activity of bisnordihydrotoxiferine, a tertiary indole alkaloid, in rodents. *Phytother. Res.*, **6**(2), 84–88.

Anti-emetic Testing

Antiemetic agents. Chapter 30 in *Screening Methods in Pharmacology—see* standard textbook list, Appendix I (for compulsive gnawing test and pigeon pecking response).

Bradley, P. *et al* (1986). Proposals for the classification and nomenclature of functional receptors for 5-hydroxytryptamine. *Neuropharmacology*, **25**, 563–576.

Bunce, K. *et al* (1991). Clinical evaluation of 5-HT$_3$ receptor antagonists as antiemetics. *Trends Pharm. Sci.*, **12**(2), 46–48.

Fozord, J.R. (1987). 5-HT$_3$ receptors and cytotoxic drug-induced vomiting (and references therein). *Trends Pharm. Sci.*, **8**, 44–45.

Ison, P.J. (1986). Neurotransmitter receptor binding studies predict antiemetic efficacy and side effects. *Cancer Treat. Rep.*, **70**, 637–641.

Anti-ulcer Testing

Allen, A. *et al* (1980). Mucus and bicarbonate secretion and their possible role in mucosal protection. *Gut*, **21**, 249–252.

Best, R. *et al* (1984). The anti-ulcerogenic activity of the unripe plantain banana (*Musa sp.*). *Br. J. Pharmacol.*, **82**, 107–116.

Croft, D.N. (1965). The estimation of deoxyribonucleic acid in the presence of sialic acid: application to analysis of gastric washings. *Biochem. J.*, **95**, 612–615.

De Soldato, P. *et al* (1985). Comparison of the gastric cytoprotective properties of atropine, ranitidine and PGE$_2$ in rats. *Eur. J. Pharmacol.*, **106**, 53–57.

Ghosal, S. *et al* (1988). Anti-ulcerogenic activity of fulvic acids and 4′-methoxy-6-carbomethoxybiphenyl isolated from Shilajit. *Phytother. Res.*, **2**(4), 187–191.

Robert, A. *et al* (1979). Cytoprotection by prostaglandins in rats. Prevention of gastric necrosis produced by alcohol, HCl, NaOH, hypertonic NaCl and thermal injury. *Gastroenterology,* **77,** 433–438.

Shay, H. *et al* (1945). Simple method for the uniform production of gastric ulceration in the rat. *Gastroenterology,* **5,** 43–47.

5

The Liver and Biliary System

The liver is the largest solid organ in the body and one of the most complex, with a wide range of functions including storage of nutrients, maintenance of carbohydrate homeostasis, secretory and excretory functions, protein synthesis and certain vital metabolic functions unique to the liver. These include aspects of hormone metabolism (insulin, glucagon, thyroxine, glucocorticoids); lipid metabolism (cholesterol, triglyceride, very low density lipoprotein (VLDL); protein metabolism (amino acid degradation, synthesis of clotting factors, urea and plasma proteins) and detoxification of xenobiotics. The gall-bladder produces bile salts necessary for lipid solubilisation and transport and together these organs may be referred to as the hepatobiliary system.

As the liver is so complex, its disorders are equally so, and therefore finding a system for testing plants for the treatment of such diseases is not a simple matter. Such testing may be carried out by causing liver damage in experimental animals and estimating beneficial effects of treatment as measured by liver function tests (enzyme levels, hexobarbital sleeping time, etc.) or by removing part of the liver and measuring the rate of regeneration. The liver has remarkable regenerative powers: up to 90% surgical resection can be followed by complete recovery. Patients with liver disorders are usually given supportive treatment (diet, removal of toxins, e.g. in drug overdosage, as with paracetamol, alcohol) rather than active therapy.

Conventional medicine does not provide many remedies for hepatitis, cirrhosis, liver damage by toxins or for biliary tract disorders. However, plant drugs in Western and traditional Chinese herbal medicine have long been used for liver and biliary disease and recent pharmacological and clinical experiments have shown them to be beneficial, as measured by standard liver function tests (see examples given). The most spectacularly successful must be silymarin, an extract of the milk (or Marian) thistle, *Silybum marianum* (Compositae), which can be used to treat almost all types of liver disease and, in particular, severe liver toxicity due to ingestion of the death cap mushroom, *Amanita phalloides*, which has a fatality rate of over 50%. Treatment with silymarin, or the principal component silibinin, within 48 h of ingestion of the fungus usually guarantees a satisfactory clinical outcome. (After the crucial 48 h liver damage leading to coma is likely, although the plant extract is still beneficial and can prevent further destruction of liver cells.) For review see Morazzoni, P. *et al* (1995), *Silybum marianum* (*Carduus marianus*). *Fitoterapia*, 66(1), 3–42.

Other plants with demonstrable hepatoprotective activity *in vitro* include liquorice, *Glycyrrhiza glabra* (Leguminosae); bitter kola, *Garcinia kola* (Guttiferae); and *Picrorhiza kurroa* (Scrophulariaceae).

Traditional oriental medicine uses several plants for the treatment of liver dysfunction, including *Plantago asiatica* (Plantaginaceae), *Gentiana scabra* (Gentianaceae), *Alisma orientale* (Alismataceae), *Polygonatum japonicum* (Liliaceae) and others. All these plants have been shown to be active in screening tests for liver protective activity using carbon tetrachloride-induced hepatotoxicity in mice.

In herbal medicine, many of the plants described as being used for liver disorders actually affect the gall-bladder (e.g. wormwood, *Artemisia absinthium*; greater celandine, *Chelidonium majus*; turmeric, *Curcuma longa*), increasing the flow of bile, and will be discussed in the relevant section. Some, e.g. *Picrorhiza kurroa*, have been shown to act on both the liver and the biliary system.

TESTING FOR LIVER PROTECTIVE ACTIVITY

The main causes of liver disease are drug or toxin-induced damage (including alcohol), viral infections and immunogenic reactions.

Obviously, during testing, liver damage has to be induced in some way and the best methods are those mimicking natural causes. These are usually chemical (e.g. paracetamol or carbon tetrachloride) or immunological (e.g. complement-mediated cytotoxicity induced by immunisation with specific antigens). Animals can then be treated or pretreated with plant extracts and the results determined by either assessing liver function, i.e. measuring enzyme levels or a parameter which is affected by these (e.g. hexobarbital hypnosis).

For screening purposes, an *in vitro* method using cultured hepatocytes may be more convenient, allowing more samples to be processed, being less costly regarding animals, and giving more reproducible results. In this assay, liver cells are isolated using collagenase and viability determined. They are precultured for a time before adding hepatotoxin and the test sample, cultured for a suitable period and then liver function assessed by measuring transaminase activity. For details *see* Kiso, Y. *et al* (1983a, 1983b, 1983c).

A method for testing the ability of plant essential oils to stimulate liver regeneration has been described, using partially hepatectomised rats, which measures the weight of liver relative to body weight before and after treatment with the oils. For details *see* Gershbeim, L. (1977), *Food Cosmet. Toxicol.*, **15**, 173–181.

Animal Models: Hepatoprotection

Either mice or rats may be used, and the test extract given to the animals either before or after treatment with the hepatotoxin depending on whether prevention or treatment is being investigated.

Chemically Induced Liver Damage

Paracetamol

Paracetamol causes liver cell damage in a dose-dependent way. If taken in overdosage, liver failure may occur within 2–3 days due to depletion of glutathione levels (the conjugate of the drug with glutathione is harmless). Paracetamol hepatotoxicity has been shown to be mediated by an electrophilic reactive metabolite (N-acetyl, p-benzoquinone imine) which covalently binds to the thiol groups of proteins and other macromolecules. Since it is a commonplace drug and responsible for many cases of poisoning, paracetamol is often used to induce hepatotoxicity during testing. Clinical treatment of paracetamol overdose is by administering N-acetyl cysteine to increase glutathione levels, and is effective only if given early enough (about 10 h after ingestion). In animals, extracts of *Schisandra* (*Schizandra*) *chinensis* fruits and *Picrorhiza kurroa* roots have been shown to protect against paracetamol-induced damage (*see* papers used as examples in results section and 'Advances in Chinese medicinal materials research' reference in 'Further Reading' list).

Carbon Tetrachloride (CCl$_4$)

Carbon tetrachloride also causes acute liver cell damage in a dose-dependent manner. It is a common solvent, although used less now that its toxicity has been realised. Damage occurs due to activation of CCl$_4$ by hepatic cytochrome P-450 to free radicals, which induce lipid peroxidation, covalent binding to macromolecules and inhibition of the calcium pump of microsomes, leading to liver cell necrosis. It is the most frequently used hepatotoxin for screening tests and is usually used with mice, since it involves lower costs and smaller samples. Plant extracts shown to be effective in preventing CCl$_4$ toxicity include *Silybum marianum*, *Picrorhiza kurroa*, *Plantago asiatica*, *Cyperus rotundus* and others.

Other Chemical Hepatotoxins

D-galactosamine is considered to produce liver damage in rats rather like that of viral hepatitis in humans from a morphological and functional viewpoint. Thioacetamide, peroxides and ionophore A23187 are also used (*see* Scheissel *et al* (1984) in Further Reading list).

Complement-mediated Hepatotoxicity

This is an attempt to produce an animal model of hepatitis similar to that produced in man by viral infection. In certain types of acute viral hepatitis, transformation to chronic hepatitis occurs and is thought to involve immune responses of the host rather than direct action of virus. Experimental hepatitis can be produced in rabbits by immunisation with human liver specific lipoprotein (LSP). The procedure will not be described here but is an *in vitro* assay method using complement-mediated cytotoxicity in primary cultured mouse hepatocytes. Damage to the hepatocytes is monitored by determination of enzyme levels. In the reference given, silybin, cynarin, glycyrrhizin, glycyrrhetinic acid, picrosides I and II were shown to exert liver protective effects: *see* Kiso, Y. *et al* (1987).

Materials and Methods: Hepatoprotection (Tables 5.1–5.8)

IN VIVO METHODS

The basic method applies regardless of hepatotoxin employed.

Animals

A group should consist of three or more animals, either mice or rats.

Control groups
(a) Solvent without hepatotoxin only.
(b) Hepatotoxin only.
(c) Test plant material only.
Test group   Hepatotoxin plus test plant extract.
Standard reference group, if required   Administer a known liver protectant such as silymarin along with hepatotoxin.

Reagents

Hepatotoxins
CCl_4, dose range: around 1/8 (mouse), 1/16 (rat) of the LD_{50}.
In mice a suitable range would be 4–8 µl/kg, given i.p. as a 40% solution in olive oil (20% for rats).

Paracetamol (Acetaminophen), dose range: around 2 g/kg body weight (rat), 500 mg/kg (mouse) in 20 ml gum acacia or methylcellulose (1%) suspension, administered orally (p.o.) using a stomach tube.

Standard liver protectant

Silymarin (50 mg/kg mice, 35 mg/kg rats) in 1% methylcellulose solution, administered intraperitoneally (i.p.).

Test solutions

Plant extract dissolved or suspended in 1% methylcellulose or gum acacia, p.o.; or in normal saline (0.9%) and injected i.p.

Method

Animals are usually allowed free access to food and water until administration of hepatotoxin, and then restricted to water only. When the experiment is due to last over 24 h, food is usually allowed again after 24 h. If preventative activity is being investigated, administer test plant solutions (usually twice) in the 24 h prior to giving hepatotoxin. Then give solvent control, hepatotoxin, and standard reference substance to relevant groups of animals. If curative activity is being studied, administer hepatotoxin to induce liver damage prior to test plant extract. Where enzyme activity is being measured, blood is taken 24 h after CCl_4 injection and 24 and/or 48 h after paracetamol ingestion. For histopathological investigation, a similar timescale may be used. For serum transaminase measurement, animals may be anaesthetised (with ketamine or a barbiturate) and blood collected from the vena cava or other suitable site, into heparinised vials. For measuring other parameters, e.g. fatty acids, peroxides and histopathological changes to the liver, animals are sacrificed by decapitation and the liver excised.

Assessment of Liver Damage

Enzyme Levels   This is the easiest and commonest method of estimating liver damage. It is possible to use several enzymes; however, those usually employed are the aminotransferases (= transaminases): *aspartate aminotransferase (AsAT)* (formerly glutamic oxaloacetic transaminase, GOT) and *alanine aminotransferase (ALAT)* (formerly glutamic pyruvic transaminase, GPT). In the older literature, serum levels of these are often represented as SGOT and SGPT respectively.

AsAT is present in liver, heart, muscle, kidney and brain, and catalyses the conversion of aspartate to oxaloacetate and glutamate. Necrosis or membrane damage releases the enzyme into circulation, therefore it can be measured in serum without resorting to taking liver samples. High levels of AsAT therefore indicate liver damage, such as that due to viral hepatitis and toxicity, as well as cardiac infarction and muscle injury.

ALAT catalyses the conversion of alanine to pyruvate and glutamate and is released in a similar manner, but is more specific to the liver.

High levels of these enzymes indicate acute hepatitis or damage, but in a

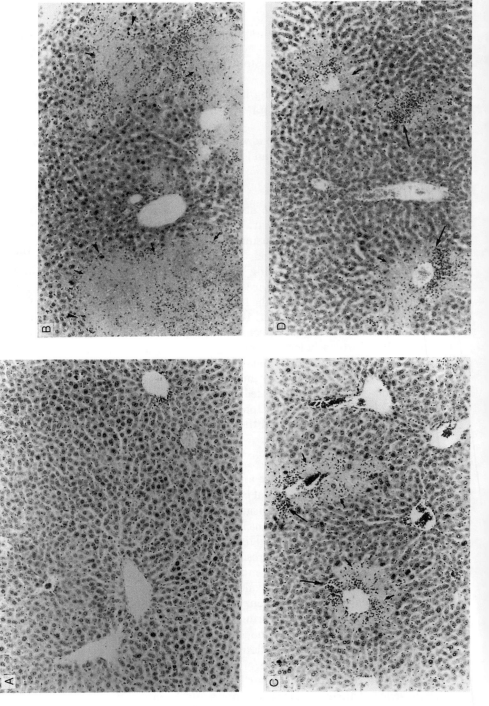

clinical situation normal levels are found in very advanced disease and alcoholic hepatitis.

Other enzymes may be used, such as: lactic dehydrogenase, γ-glutamyl transpeptidase, alkaline phosphatase, glucose-6-phosphatase. These are normally used in conjunction with AsAT and ALAT determinations.

Plasma is obtained by centrifugation of blood ($2-3000 \times g$ for 5 min). All enzyme determinations are best carried out using a commercially obtained kit (e.g. from Boerhinger, Darmstadt, Germany; Wako Pure Chemical Industries, Osaka, Japan) and following the instructions.

Modification of Barbiturate-induced Sleep   The principle behind this method is that if liver microsomal enzymes have been destroyed by disease or a hepato-toxin, the metabolism and excretion of a barbiturate (usually hexobarbital) will be impaired, and therefore hypnosis will be prolonged compared to that of a control group. The advantages of this method are that very little in the way of equipment or reagents is needed, since all that is required is for all groups to be subsequently injected i.p. with hexobarbital, and sleeping times measured. It is, however, expensive regarding animals: mice are usually used and groups need to be larger (about 10 mice).

A suitable protocol is given below:

Day	Control	Toxin (CCl_4)	Test
1	Saline	Saline	Extract
2	Saline	CCl_4	Extract + CCl_4
3	Saline	CCl_4	Extract + CCl_4
4	Saline	Saline	Extract
5	Hexabarbital	Hexabarbital	Hexabarbital

Histopathological Studies of Liver Damage (Figure 5.1)   This gives a visual assessment of the damage but is not suitable for screening or accurate quanti-tative analysis. Slices of liver are taken, usually on the fourth day after admin-istration of the toxin, fixed in formalin (10%), embedded in paraffin wax and cut into $4-5\,\mu m$ sections. The deparaffinised sections are stained with haema-toxylin and eosin and viewed with a light microscope at a magnification of $\times 30$ or higher. Examples of photographs so obtained are given in Figure 5.1.

Figure 5.1   Liver sections taken from mice. (a) The control group (corn oil only). Normal appearance of the liver architecture. (b) The CCl_4-treated group. Patch necrosis (→) with mild inflammation. Some pyknotic cells around the lesion (▶). (c) The *Bupleurum chinense*-treated group (CCl_4 + *B. chinense*). Collar necrosis (→) with mild to moderate inflammation (——→). Portal tract is not remarkable. (d) The *Eclipta prostrata*-treated group (CCl_4 + *E. prostrata*). Collar necrosis (→) with moderate inflammatory cell (——→) of central veins. From Hepatoprotective activity of Taiwan Folk Medicine: *Eclipta prostrata* against hepatotoxicity induced by various toxins. Lin, S.-C. *et al* (1996), *Phytotherapy Research* **10** (in press).

Table 5.1 Influence of various doses of CCl_4 on ALAT plasma levels in mice, 24 h after intoxication

CCl_4 ($\mu l/kg$)	Number of mice	ALAT (U/l)
0	21	21 (16–27)
2	3	54 (42–69)[a]
4	3	76 (45–128)[a]
7	4	183 (157–213)[a]
8	3	487 (403–586)[a]

[a] $p < 0.01$ As compared with placebo group. From Study of antihepatotoxicity of *Eupatorium cannabium* L., Lexa, A., Fleurentin, J., Younos, C. and Mortier, F. (1990), *Phytother. Res.*, 4(4), 148–151, with permission.

Table 5.2 Influence of a pretreatment of *E. cannabinum* and silymarin on ALAT plasma levels in mice after CCl_4 intoxication

Treatment	Dose (mg/kg)	CCl_4 ($\mu l/kg$)	Number of mice	ALAT (IU/l)	Protection (%)
Control without CCl_4	–	0	21	21 (16–27)	
Placebo	–	7	14	160 (121–212)	–
E. cannabinum	62.5	7	10	88 (37–180)	51
	125	7	10	81 (65–268)	56
	250	7	10	131 (65–268)	21
Placebo	–	7	14	139 (74–261)	
E. cannabinum	500	7	10	36 (33–93)[a]	70
Placebo	–	7	20	122 (69–252)	
E. cannabinum	1000	7	10	36 (19–68)[a]	84
Placebo	–	7	15	165 (123–220)	–
Silymarin	–	7	–	–	–
	–	7	–	–	
	100	7	10	41 (20–86)[b]	86

[a] $p \leqslant 0.001$ As compared with placebo group. [b] $p \leqslant 0.01$ As compared with placebo group. From Study of antihepatotoxicity of *Eupatorium cannabium* L., Lexa, A., Fleurentin, J., Younos, C. and Mortier, F. (1990), *Phytother. Res.*, 4(4), 148–151, with permission.

Treatment of Results

Testing for Antihepatotoxicity

Hepatotoxicity Test Adjustment   All the CCl_4 doses induced a significant increase in plasma ALAT levels relative to the control group (Table 5.1). To obtain data of the same magnitude as those observed in an hepatotoxicity test performed using rats in our laboratory (Lexa *et al.*, 1989), the dose of $7 \mu l/kg$ was chosen for the later experiments.

E. cannabinum Extract Pretreatment   *E. cannabinum* pretreatment induced a significant protective trend at 500 and 1000 mg/kg (70 and 84%, respectively). Decreasing doses (250, 125 and 62.5 mg/kg) were inefficient (21%, 56% and 51%, respectively). Silymarin pretreatment induced a significant decrease in plasma ALAT relative to the placebo group with 100 mg/kg dose (86% of protection) (Table 5.2). Eupatoriopicrin (5 mg/kg) pretreatment was

Table 5.3 Influence of a pretreatment of eupatoriopicrin on ALAT plasmatic levels in mice after CCl_4 intoxication

Treatment	Dose (mg/kg)	CCl_4 (μl/kg)	Number of mice	ALAT (IU/l)
Control (ethanol 10%)	–	0	10	16 (14–19)
Placebo (ethanol 10%)	–	7	10	212 (194–278)
Eupatoriopicrin	5	7	10	206 (172–244)

From Study of antihepatotoxicity of *Eupatorium cannabium* L., Lexa, A., Fleurentin, J., Younos, C. and Mortier, F. (1990), *Phytother. Res.*, 4(4), 148–151, with permission.

Table 5.4 Influence of a post-treatment of *E. cannabinum* extract on ALAT plasma levels in mice after CCl_4 intoxication

Treatment	Dose (mg/kg)	Number of mice	ALAT (24 h) (IU/l)	Protection (%)
Placebo	–	8	115 (57–235)	–
E. cannabinum	1000	10	85 (41–176)	33

From Study of antihepatotoxicity of *Eupatorium cannabium* L., Lexa, A., Fleurentin, J., Younos, C. and Mortier, F. (1990), *Phytother. Res.*, 4(4), 148–151, with permission.

ineffective. No toxicity or potentiation of CCl_4 hepatotoxicity was noted with the solvent injection and with eupatoriopicrin (Table 5.3).

E. cannabinum Extract Post-treatment The aqueous extract (1000 mg/kg) administered i.p. 30 min after CCl_4 was very slightly (33% of protection) and not significantly protective against ALAT leakage (Table 5.4).

Changes Induced by Paracetamol

Table 5.5 Effect of paracetamol on normal rat serum

Parameters	Control (Group I)	Time after paracetamol treatment (2 g/kg body wt)	
		24 h (Group II)	48 h (Group III)
Glutamate oxaloacetate transaminase[a]	131.13 ± 6.65	198.34 ± 12.77	216.89 ± 16.89
Glutamate pyruvate transaminase[a]	52.00 ± 6.67	234.20 ± 29.18	261.36 ± 23.58
Alkaline phosphatase[b]	21.41 ± 2.48	33.50 ± 5.82	79.65 ± 6.54
Bilirubin[c]	0.41 ± 0.08	0.76 ± 0.17	1.44 ± 0.21

[a] μmol of pyruvate formed/min/l serum.
[b] μmol of *p*-nitrophenol released/min/dl serum.
[c] mg/dl serum.
Values are mean ± SD from 6 animals. $p < 0.01$, as compared to control (Group I). From Prevention of paracetamol-induced hepatic damage in rats by picroliv, the standardised active fraction from *Picrorhiza kurroa*. Dwivedi, Y., Rastogi, R., Garg, N. and Dhawan, B. (1991), *Phytother. Res.*, 5, 115–119, with permission.

Effect of Picroliv on Paracetamol-induced Biochemical Changes The serum values summarised in Table 5.6 indicate that Picroliv produced 80–100% reversal in the elevated levels of transaminases and 70–80% protection against the increase in alkaline phosphatase activity. The increase in serum bilirubin was also significantly reversed (62% at 6 mg/kg and 100% at 12 mg/kg dose of Picroliv).

Table 5.6   Effect of Picroliv on serum of paracetamol-treated rats

Parameters	Picroliv[a]	
	6 mg/kg body wt	12 mg/kg body wt
Glutamate oxaloacetate transaminase	125.38 ± 3.78	130.65 ± 3.33
Glutamate pyruvate transaminase	93.70 ± 5.37	58.54 ± 4.13
Alkaline phosphatase	39.86 ± 3.64	31.82 ± 2.62
Bilirubin	0.80 ± 0.03	0.39 ± 0.03

All values expressed as in Table 5.5.
[a] After 48 h of paracetamol (2 g/kg) administration.
Values are mean ± SD from 6 animals and differ significantly ($p < 0.01$) from the corresponding values of Group III in Table 5.5. From Prevention of paracetamol-induced hepatic damage in rats by picroliv, the standardised active fraction from *Picrorhiza kurroa*. Dwivedi, Y., Rastogi, R., Garg, N. and Dhawan, B. (1991), *Phytother. Res.*, **5**, 115–119, with permission.

Pretreatment with ethanol extract of Withania frutescens *leaf or silymarin* When the ethanol extract of *Withania frutescens* leaf was administered 24 h before CCl_4 at a dose equivalent to 1/4 of its LD_{50}, the extract significantly reduced mortality, as reflected by the 87% rise in the LD_{50}. Silymarin was not active at the dose used (50 mg/kg) (Table 5.7). The increase caused by CCl_4 in barbiturate-induced sleep was clearly reduced when *W. frutescens* extract was given prophylactically (Table 5.7). This reduction was greater than that obtained with silymarin. Bile output recovered almost completely when plant extract or silymarin was given, in comparison to the decline in flow caused by CCl_4.

Levels of SGOT and SGPT, which rose after CCl_4 was administered, declined significantly in animals treated previously with *W. frutescens* extract or silymarin, in comparison with the values found in rats given CCl_4 alone (Table 5.7). The decrease in SGOT levels caused by the plant extract, however, was greater than that obtained with silymarin, a finding which suggests that hepatocellular activity by the former is more intense. The recovery of normal values of arachidonic acid and oleic acid when *W. frutescens* extract was administered 24 h before CCl_4 may indicate that the extract limits the peroxidative effect of carbon tetrachloride. This effect was, however, not similar to that of silymarin in all cases (Table 5.7).

Table 5.7 Influence of pretreatment with the ethanol extract of *Withania frutescens* leaf or silymarin on the LD_{50} of CCl_4, Nembutal-induced sleep time, SGOT, SGPT, oleic acid and arachidonic acid after CCl_4 intoxication

Parameter	Treatment[a]			
	I	II	III	IV
Mice				
LD_{50} CCl_4 (ml/kg)		2.1 ± 0.32	3.9 ± 0.21	1.9 ± 0.24
Nembutal (50 mg/kg) sleeping time (min)	61.0 ± 6.68	348.0 ± 85.83	151.0 ± 26.62[b]	228.0 ± 41.68[b]
Rats				
SGOT (IU/l)	51.0 ± 11.98	1832.5 ± 311.02	339.0 ± 72.14[b]	749.0 ± 250.39[b]
SGPT (IU/l)	13.5 ± 3.92	760.0 ± 212.52	139.2 ± 19.00[b]	265.5 ± 105.06[b]
Oleic acid (%)	20.2 ± 3.19	14.8 ± 1.68	18.9 ± 2.13[b]	20.58 ± 1.27[b]
Arachidonic acid (%)	20.6 ± 0.66	17.1 ± 2.72	20.5 ± 1.90[c]	17.9 ± 1.05

[a] I, Control; II, animals given CCl_4 alone; III and IV, animals treated with ethanol extract of *W. frutescens* leaf or silymarin, respectively, 24 h prior to the administration of CCl_4.
[b] $p < 0.01$ mean values $\pm$ SE as compared with CCl_4.
[c] $p < 0.05$ mean values $\pm$ SE as compared with CCl_4.
From The protective and curative action of *Withania frutescens* leaf extract against CCl_4-induced hepatotoxicity, Montilla, M. *et al* (1990), *Phytother. Res.*, 4(6), 212–215, with permission.

Post-treatment with ethanol extract of W. frutescens *leaf or silymarin* The duration of Nembutal-induced sleep potentiated by CCl_4 was clearly reduced by the *W. frutescens* extract (Table 5.8), thus suggesting that the extract acts by restoring the diminished enzymatic capacity of the liver. Silymarin at the dose used in this study had no effect on barbiturate-induced sleep.

TESTING FOR CHOLERETIC AND ANTICHOLESTATIC ACTIVITY

Disturbed function of the biliary system often leads to intolerance of fats, producing indefinite symptoms including pain and discomfort in the epigastrium, particularly after a rich meal. Symptoms usually improve with a low fat diet.

Table 5.8 Influence of a post-treatment with the ethanol extract of *W. frutescens* leaf or silymarin on the Nembutal-induced sleep time in mice

Parameter	Treatment[a]			
	I	II	III	IV
Nembutal (50 mg/kg) sleeping time (min)	61.0 ± 6.68	245.0 ± 33.63	172.0 ± 43.56[b]	292.0 ± 28.15[b]

[a] I, Control; II, animals given CCl_4 alone; III and IV, animals treated with ethanol extract of *W. frutescens* leaf or silymarin for 3 consecutive days after the administration of CCl_4.
[b] $p < 0.05$ mean values $\pm$ SE as compared with CCl_4.
From The protective and curative action of *Withania frutescens* leaf extract against CCl_4-induced hepatotoxicity, Montilla, M. *et al* (1990), *Phytother. Res.*, 4(6), 212–215, with permission.

A failure in the mechanics of bile flow is termed biliary dyskinesia and can be treated with plant drugs, the efficacy of which has been shown *in vivo* by methods which will be described. Cholestasis is a cessation in bile flow and is very serious if prolonged. It can have many causes, including gall-stones, viral hepatitis, pancreatitis, tumours, and drugs such as alcohol, methyl testosterone, oestrogens, chlorpromazine and other phenothiazines.

Substances which promote an increase in the production of bile are sometimes termed cholagogues, and those which increase the flow of bile are choleretics; however, the division is theoretical and in practice both effects may be termed choleretic. Anticholestatics are those which reverse or prevent cholestasis.

Bile acids are produced by the liver, secreted from the canaliculi and through the ductules into the gall-bladder. Contraction of the gall-bladder occurs in response to the release of the hormone cholecystokinin in the upper small bowel. The primary bile acids, cholic and chenodeoxycholic acids, are synthesised from cholesterol and excreted into the bile and thence the alimentary canal. These can be conjugated with taurine and glycine to form bile salts. In the colon, secondary bile acids are produced by 7-α-dehydroxylation to form deoxycholic and lithocholic acids. The function of bile salts is lipid solubilisation, both of cholesterol and lecithin in the bile itself, and also dietary lipids prior to absorption. This is possible because of the detergent properties enabling micelle formation.

Gall-stones usually form in the gall-bladder and may or may not migrate to the common bile duct. Although often asymptomatic, they may obstruct the cystic duct and, if followed by infection, cause acute cholecystitis; or by passing into the common bile duct cause biliary colic, jaundice or cholangitis. They may be composed largely of cholesterol, with some calcium salts (radiolucent unless sufficient calcium to render radio-opaque), or bilirubin with phosphate or carbonate, known as 'pigment stones'. Cholesterol stones are occasionally treated with oral supplementary bile acids, but the usual treatment is surgery (cholecystectomy).

Some important plant remedies used in different parts of the world to treat gall-bladder disorders include wormwood and oriental wormwood, *Artemisia absinthium* and *A. scoparia*; celandine, *Chelidonium majus*; kalmegh, *Andrographis paniculata*; and kutki, *Picrorhiza kurroa*. These have been shown to increase the flow of bile and may also change the composition of bile acids and physical properties of bile.

Obviously these experiments are done *in vivo*, using anaesthetised rats or guinea-pigs, and tests may be carried out for choleretic activity, a simple increase in the flow of bile; and for anticholestatic activity, where bile flow has been reduced by the use of drugs such as ethinyloestradiol or paracetamol. It is necessary also to test the bile so produced, even if only by simple gravimetric analysis, to ensure that the drug is not merely stimulating aqueous secretion into the bile.

Animal Models: Choleretic Activity

Either rats or guinea-pigs may be used, since they have bile ducts large enough to work with. Plant extracts are administered to groups of animals either orally (p.o.) or by intraperitoneal injection (i.p.), and the effect on either bile flow or experimentally induced cholestasis then studied.

Choleretic effects can be measured on conscious animals, usually rats, if a cannula has previously been inserted into the common bile duct and passed beneath the skin and out at a suitable point. Bile can then be collected into a graduated tube. Measurements may also be carried out in guinea-pigs whilst kept under anaesthetic; both will be described in the 'Methods' section.

If anticholestatic effects are to be studied, a cholestatic agent such as para-cetamol or ethinyloestradiol is administered either after or during dosage with the plant extract, and bile flow measured in the animal after cannula-tion in a similar manner. Ethinyloestradiol, like other sex hormones, causes biliary obstruction known as 'pure cholestasis', whereas paracetamol causes cholestasis together with liver cell damage described as 'cholestatic hepatitis'. It is obviously useful to employ different cholestatic agents to get some idea whether the action is on the biliary system directly or by preventing liver damage. Other agents which have been used include nithiocyamine, a new antischistosomiasis drug with known hepatobiliary toxicity: *see* Liu, C.-X. and Ye, G.-Z. (1991).

Bile is measured as the volume of flow produced over a particular time, usually several hours, and total solids, bile acids and salts estimated and compared with the results obtained from control groups of animals.

Materials and Methods: Choleretic and Anticholestatic Activity (Tables 5.9–5.12)

Animals

A group should consist of about eight animals, either adult rats or guinea-pigs, of either sex.

Testing for choleretic activity only
Control group
(a) Untreated animals.
Test group
(b) Test plant material only.

Testing for anticholestatic activity
Control groups
(a) As above.
(b) As above.

(c)   Cholestatic agent only.
Test group
(d)   Test plant extract plus cholestatic agent.
Standard reference group, if required   Administer a known liver protectant or anticholestatic agent such as silymarin prior to treating with cholestatic agent.

Reagents

Cholestatic agents
Ethinyloestradiol (EE): 5 mg/kg body weight (rat) injected subcutaneously daily for 3 days.
Paracetamol: a single p.o. dose 1.5 g/kg (rat and guinea-pig) produces marked hepatic damage.
Standard anticholestatic agent
Silymarin 12–20 mg/kg (rat and guinea-pig) administered p.o.
Test solutions
Varying concentrations in aqueous solution or suspension, usually administered p.o.

Method

For simple choleretic measurements, plant extracts are administered daily to test groups for about a week, then bile flow compared with that of the untreated control group. For anticholestatic testing, plant extracts are administered again for about a week, with cholestasis being induced either after the last dose (in the case of paracetamol) or for the last three days (in the case of ethinyloestradiol). Bile flow measurements are taken 48 h after paracetamol treatment or 24 h after the last dose of ethinyloestradiol.

Collection of bile   Cannulation may be carried out as follows (methods according to Shukla *et al* (1991):

1.   *Bile flow in conscious rats.*   The abdomen is opened whilst under ether anaesthesia, the common bile duct exposed and cannulated with polyethylene tubing just before the bifurcation. After cannulation the abdominal opening is stitched and a small cut made on the back near the head for outlet of tubing, which passes beneath the skin and thence out of the animal, and the skin is restitched. The tubing is then passed through a flexible steel spring to allow free movement of the animal, and out of the cage for collection of bile into a graduated tube or a series of small pre-weighed tubes. This may be carried out over 24 h or as long as required. This method has the advantage of using animals in as natural a state as is possible under laboratory conditions.

2. *Bile flow in anaesthetised guinea-pigs.* Guinea-pigs may be anaesthetised with urethane (0.6 ml/100 g of a 25% solution), the abdomen opened and the bile duct cannulated by polyethylene tubing. Bile is again collected into a graduated tube. This may be done over about 5 h.

3. *Bile flow in anaesthetised rats* may be studied using the method of Tripathi, G. and Tripathi, Y. (1991), in which the animals are anaesthetised with pentobarbitone sodium (30 mg/kg) and bile collected via a cannula (polyethylene, No. 48) in a similar way.

Estimation of bile contents (Methods recommended by Prof. B.N. Dhawan) Simple estimation of total solids may be carried out by drying measured volumes of bile to constant weight in a vacuum oven at ambient temperature. This is to show that a true difference in production of bile is obtained, rather than a dilution or other effect. For better comparisons, a more detailed analysis of bile is usually carried out.

- *Bile salts.* (Method of Hawk, P.B. *et al* (1954); *see* 'Further Reading'.)
  Take 25 ml of undiluted bile in an evaporating dish and add enough charcoal to form a paste. Evaporate to dryness on a water bath. Remove the residue, grind in a mortar and transfer to a small flask. Add about 50 ml absolute alcohol and boil on a water bath for 20 min. Filter and add solvent ether to the filtrate until there is a slight permanent cloudiness. Cover the vessel and set aside until crystallisation is complete. Evaporate and weigh the resultant bile salt crystals.

- *Bile acids.* (Method of Mosbach, E.H. *et al* (1954).)
  Take an aliquot (about 1 ml) of the bile sample in a 50 ml Erlenmeyer flask and add 5 ml ethanol. Heat to boiling on a steam bath, and filter into a 25 ml stoppered centrifuge tube. Use 2 portions (of about 5 ml) of hot ethanol to rinse the flask and filter paper, and add to the flask. Evaporate to dryness on a steam bath under a current of air. Add 5 ml of a 5% solution of sodium hydroxide and hydrolyse (e.g. by autoclaving for 3 h). Cool to room temperature and acidify with conc. HCl. Then extract with 4 portions of diethyl ether, combine the extracts, back-wash with distilled water and dry over sodium sulphate for 2–3 h. Filter (after back-washing the sodium sulphate with 5 ml ether) and evaporate on a water bath. Dissolve the residue in 10 ml acetone, pipette an aliquot (usually 1 ml) into a 25 ml centrifuge tube and evaporate off the acetone. Add 5 ml sulphuric acid (65% and heat to about 60 °C on a water bath for 5 min. Allow to cool (using tap water) for 15 min and measure UV absorption at 320 nm and 385 nm.

Both obey the Beer–Lambert law at these wavelengths and the optical densities remain constant for at least an hour after cooling. Compare with a standard curve of optical density vs. concentration, using test solutions of cholic and deoxycholic acids (e.g. 0.02–0.1 mg in 5 ml) in 65% H_2SO_4.

Statistical analysis

Student's *t*-test is used to compare results of the groups and expressed as the mean ± standard deviation (SD).

Treatment of Results

Choleretic Effects

See Table 5.9.

Picroliv In anaesthetised guinea-pigs, significant increases in bile flow (56.6 and 132%), bile salts (48.6 and 100%), cholic acid (67.4 and 97%), and deoxycholic acid (41.4 and 73.4%) were observed with Picroliv in doses of 6 and 12 mg/kg. The lower doses of 1.5 and 3 mg/kg did not possess significant choleretic effects.

Silymarin In anaesthetised guinea-pigs, silymarin also showed marked increases in all the parameters at the higher doses of 12 and 20 mg/kg (44 and 52% in bile flow, 64 and 79% in bile salt, 98.3 and 125% in cholic acid, 41.4 and 150% in deoxycholic acid). Some effects (22.2 and 20.6%) were noticed in bile flow and bile salts at the lower dose of 6 mg/kg along with a significant increase (75.9%) in cholic acid but no effect on deoxycholic acid. A lower dose (3 mg/kg) was not active.

Table 5.9 Choleretic effects of Picroliv and silymarin in anaesthetised guinea-pigs

	Parameter	Normal	Picroliv (mg/kg × 7)				Silymarin (mg/kg × 7)		
			1.5	3	6	12	6	12	20
1.	Bile flow (ml/100 g/h)	0.54 ± 0.02	0.55 ± 0.03	0.61* ± 0.05	0.85** ± 0.09	1.27** ± 0.13	0.66*** ± 0.02	0.78** ± 0.02	0.82** ± 0.06
2.	Bile salts (mg/ml)	1.89 ± 0.01	1.70 ± 0.03	1.88 ± 0.01	2.81** ± 0.68	3.80** ± 0.20	2.28** ± 0.02	3.10** ± 0.12	3.4** ± 0.10
3.	Bile acids (µg/ml)								
	i) Cholic acid	1.04 ± 0.01	1.06 ± 0.01	1.13* ± 0.13	1.39** ± 0.02	2.05** ± 0.01	1.83** ± 0.04	2.00** ± 0.16	2.35** ± 0.49
	ii) Deoxycholic acid	0.94 ± 0.04	0.93 ± 0.03	0.94 ± 0.01	1.33** ± 0.04	1.63** ± 0.59	0.96 ± 0.05	1.33** ± 0.55	2.35** ± 0.49

**p < 0.001; *p < 0.01.
Values mean ± SD from 8 animals each.
From Choleretic effect of picroliv, the hepatoprotective principle of *Picrorhiza kurroa*. Shukla, B. *et al* (1991), *Planta Medica*, **57**, 29–33, with permission.

Effect of Andrographolide on Bile Flow Rate   Twenty-four albino rats of H.M. strain (male, average body weight 200 g) were divided into four groups of six rats in each. The animals were allowed access to water and diet *ad libitum*. The first group was given an i.p. injection of distilled water and served as control and groups 2, 3 and 4 received i.p. injections and andrographolide (2% solution of carboxy methyl-cellulose) suspension at a dose of 20 mg, 40 mg and 80 mg/kg body weight. After 90 min of drug administration, bile flow was measured.

Results are given in Table 5.10. There was a significant enhancement in the rate of bile flow ($p < 0.05$) at a dose of 20 mg/100 g body weight. The effect was dose-dependent.

Table 5.10   Effect of andrographolide on bile flow rate in albino rats

Group	n	Dose of treatment (mg/kg body weight)	Bile flow rate[†] (mean ± SE)
1. Control	16	–	21.83 ± 1.27
2. Andrographolide treatment[‡]	6	20	27.01 ± 1.36
3. Andrographolide treatment[‡]	6	40	32.49 ± 1.34[a]
4. Andrographolide treatment[‡]	6	80	35.49 ± 1.51[b]

[a] $p < 0.01$; [b] $p < 0.001$.
[†] Bile flow was measured as mg/5 min/100 g body weight.
[‡] Andrographolide was suspended in 2% solution of carboxy methyl cellulose and injected i.p.
From Choleretic action of andrographolide obtained from *Andrographis paniculata* in rats. Tripathi, G.S. and Tripathi, Y.B. (1991), *Phytother. Res.*, **5**, 176–178, with permission.

Effect of Andrographolide on Physical Properties of Bile   The results (Table 5.11) indicate the increase in the solid residue in the bile compared with the control value. It seems that this drug does not merely stimulate aqueous secretion

Table 5.11   Effect of andrographolide on bile composition in albino rats

Group	n	Dose of andrographolide treatment (mg/kg body weight)	Density (mg/ml) (mean ± SE)	Bile residue (mean ± SE)	Aqueous (mean ± SE)
Control	16	–	960 ± 10.0	12.42 ± 1.54	948.33 ± 8.29
Andrographolide treatment[†]	6	20	972.92 ± 11.8	18.42 ± 0.96[a]	954.50 ± 10.43
Andrographolide treatment[†]	6	40	988.06 ± 12.2[b]	21.60 ± 0.85[b]	967.00 ± 11.17[a]
Andrographolide treatment[†]	6	80	1015.9 ± 12.0[b]	31.50 ± 1.37[b]	984.40 ± 6.58[b]

[a] $p < 0.01$; [b] $p < 0.001$.
[†] Andrographolide was suspended in 2% solution of carboxy methyl-cellulose and injected i.p.
From Choleretic action of andrographolide obtained from *Andrographis paniculata* in rats. Tripathi, G.S. and Tripathi, Y.B. (1991), *Phytother. Res.*, **5**, 176–178, with permission.

into the bile but also bile salts and other solid materials. Thus, it could be concluded that it is a true choleretic agent and the effect is dose-dependent.

Anticholestatic Effects

See Table 5.12

Paracetamol-induced Cholestasis   Paracetamol reduced the volume of bile and the amount of bile salts by 75% and 63.8%, respectively. There was a 38.3% reduction in cholic acid and 39.6% in deoxycholic acid content.

Picroliv   Treatment with Picroliv (6 and 12 mg/kg) resulted in complete reversal of the decrease in bile flow. Lower doses of 1.5 and 3 mg/kg gave 25.0% and 50.0% protection, respectively. A complete recovery of the decreased bile salt content was also noticed with the higher doses and 76.9% protection was obtained with the 3 mg/kg dose. A lower dose (1.5 mg/kg) was not effective. In cholic acid, the reversal was 45.1–81.9 with 3–12 mg/ kg doses of Picroliv. The dose of 1.5 mg/kg could exhibit only a 7.2% protection. A significant dose-dependent recovery was also seen in deoxycholic acid (36.1–100%) with Picroliv in doses of 1.5–12 mg/kg.

Silymarin   Silymarin (6–20 mg/kg) showed 41.6–91.6% recovery of the decreased volume of bile. A greater (67.6–100%) protection of bile salts was observed at the above dose levels. Dose-dependent effects of silymarin were also seen on changes in cholic acid (19.2–78.9%) and deoxycholic acid (25.5–82.9%) contents. The results are summarised in Table 5.12.

FURTHER READING

General Reading

Brunt, P.W., Lowsowsky, M.S. and Read, A.B. (1984). *The Liver and Biliary System*. Heinemann, London.
Carulli, N. *et al* (1975). Alteration of drug metabolism during cholestasis in man. *Eur. J. Clin. Invest.*, **5**, 455–462.
Hawk, P.B., Oser, B.L. and Summerson, W.H. (1954). *Practical Physiological Chemistry*. J. and A. Churchill Ltd., London.

Methods and Examples

Hepatoprotection

Chang, H.M. *et al* Ed. (1985). *Advances in Chinese Medicinal Materials Research. Session III, Liver Diseases*, pp. 205–285. World Scientific Publishing Co. Pte Ltd, Singapore.
Chaudhury, B.R. *et al* (1987). *In vivo* and *in vitro* effects of kalmegh. *Andrographis paniculata*, extract and andrographolide on hepatic microsomal drug metabolizing enzymes. *Planta Medica*, **53**(2), 135–140.

Table 5.12 Anticholestatic effect of Picroliv and silymarin against paracetamol (1.5 mg/kg *p.o.*) induced cholestasis in conscious rat

Parameter	Normal	Paracetamol	Picroliv + Paracetamol (mg/kg × 7)				Silymarin + Paracetamol (mg/kg × 7)		
			1.5	3	6	12	6	12	20
1. Bile flow (ml/100 g/h)	0.16 ± 0.06	0.04** ± 0.001	0.07** ± 0.008	0.10** ± 0.001	0.21** ± 0.02	0.20** ± 0.08	0.09** ± 0.005	0.13** ± 0.004	0.15** ± 0.02
2. Bile salts (mg/ml)	3.54 ± 0.04	1.28 ± 0.04	0.95 ± 0.05	3.02** ± 0.01	3.54** ± 0.47	4.36** ± 0.33	2.81** ± 0.08	2.93** ± 0.24	3.61** ± 0.22
3. Bile acids (µg/ml)									
i) Cholic acid	4.33 ± 0.06	2.67** ± 0.05	2.79 ± 0.17	3.42** ± 0.56	3.84** ± 0.28	4.03** ± 0.16	2.99 ± 0.13	3.78** ± 0.44	3.98** ± 0.55
ii) Deoxycholic acid	2.37 ± 0.10	1.43** ± 0.02	1.77* ± 0.33	1.97** ± 0.05	2.14** ± 0.47	3.64** ± 0.55	1.67 ± 0.47	1.99** ± 0.02	2.21** ± 0.18

** $p < 0.001$; * $p < 0.01$.
Normal group compared with paracetamol.
Values mean ± SD from groups of 8 animals each.
From Choleretic effect of Picroliv, the hepatoprotective principle of *Picrorhiza kurroa*. Shukla, B. *et al* (1991), *Planta Medica*, 57, 29–33, with permission.

Hodson, E. *et al* Eds (1980). Biochemical toxicology of acetaminophen. In *Reviews in Biochemical Toxicology*, Vol. II, pp. 103–129. Elsevier North Holland, New York.

Karmen, A. *et al* (1955). Transaminase activity in human blood. *J. Clin. Invest.*, **34**, 126–131.

Kiso, Y. *et al* (1985). Assay methods for antihepatotoxic activity using peroxide-induced cytotoxicity in primary cultured hepatocytes. *Planta Medica*, **51**, 50–52.

Kiso, Y. *et al* (1987). Assay method for antihepatotoxic activity using complement-mediated cytotoxicity in primary culture hepatocytes. *Planta Medica*, **52**(3), 241–247.

Lin, S.-C. *et al* (1944). Hepatoprotective effects of Taiwan Folk Medicine: *Alternanthera sessilis* on liver damage induced by various hepatotoxins. *Phytother. Res.*, **8**(7), 391–398.

Montgomery, R. (1957). Determination of glycogen. *Arch. Biochem. Biophys.*, **67**, 378–386.

Mori, H. *et al* (1980). Stereological analysis of Leydig cells in normal guinea-pig tests. *J. Electron Microsc.*, **29**, 8–21.

Reitman, M.D. and Frankel, P.D. (1957). A colorimetric method for the determination of serum glutamic oxaloacetic acid and glutamic pyruvate transaminases. *Am. J. Clin. Pathol.*, **28**, 56–63.

Roberts, J.C. *et al* (1987). Pro-drugs of L-cysteine as protective agents against acetaminophen-induced hepatotoxicity. 2-(Polyhydroxyalkyl)- and 2-(polyacetoxyalkyl)-thiazolidine-4(R)-carboxylic acids. *J. Med. Chem.*, **30**, 1891–1896.

Rumack, B.H. *et al* (1981). Acetaminophen overdose: 662 cases with evaluation of oral acetylcysteine treatment. *Arch. Int. Med.*, **141**, 380–385.

Schiessel, C. *et al* (1984). ^{45}Ca uptake during the transition from reversible to irreversible liver damage induced by D-galactosamine *in vivo*. *Hepatology*, **4**, 855–861.

Vogel, G. and Temme, I. (1969). Curative antagonism of liver damage caused by phalloidin with silymarin as a model of antihepatotoxin therapy. *Arzneim. Forsch.*, **19**, 613–615.

Walker, R.M. *et al* (1980). Acetaminophen-induced hepatotoxicity in mice. *Lab. Invest.*, **42**, 181–189.

Wendel, A. and Feuerstein, S. (1981). Drug-induced lipid peroxidation in mice. I. Modulation by monooxygenase activity, glutathione and selenium status. *Biochem. Pharmacol.*, **30**, 2513–2520.

Choleretic and Anti-cholestatic Activity

Berthelot, P. *et al* (1970). Mechanism of phenobarbital-induced hypercholesteresis in rat. *Am. J. Physiol.*, **219**, 809–813.

Boelsterli, U.A. *et al* (1983). Modulation by S-adenosyl-L-methionine of hepatic Na$^+$, K$^+$-ATPase, membrane fluidity and bile flow, in rats with ethinyloestradiol-induced cholestasis. *Hepatology*, **3**, 12–17.

Liu, C.-X. and Ye, G.-Z. (1991). Choleretic activity of p-hydroxyacetophenone isolated from *Artemisia scoparia* in the rat. *Phytother. Res.*, **5**, 182–184.

Mosbach, E.H. *et al* (1954). Determination of deoxycholic and cholic acids in bile. *Arch. Biochem. Biophys.*, **57**, 402–409.

Shukla *et al* (1991). Choleretic effect of Picroliv, the hepatoprotective principle of *Picrorhiza kurroa*. *Planta Medica*, **57**, 29–33.

Tripathi, G. and Tripathi, Y. (1991). Andrographolide obtained from *Andrographis paniculata* in rats. *Phytother. Res.*, **5**, 176–178.

Wannagat, F.J. *et al* (1978). Bile acid-induced increase in bile acid-independent flow and plasma membrane $Na^+ K^+ ATPase$ activity in rat liver. *J. Clin. Invest.*, **61**, 297–307.

6

The Cardiovascular System

Diseases of the cardiovascular (CV) system constitute major causes of death throughout the world. Conditions such as hypertension lead to other types of disease, such as stroke, kidney and heart disease, and need to be treated. Blood pressure is itself affected by other existing disease states such as athero-sclerosis and arrhythmias. Treatment is usually life-long; therefore drugs must be effective and safe over a long period. For detailed descriptions of types of disease, underlying aetiology and clinical management, textbooks and review articles given at the end of the chapter should be consulted. In this section the following are covered: antihypertensives, cardiotonics, anti-sclerotics and antiplatelet drugs. Diuretics, although vitally important in treating hypertension, will be covered later.

The plant kingdom already furnishes many important cardiovascular drugs. The cardiac glycosides, digoxin and lanatoside C (from *Digitalis* spp.), ouabain from *Strophanthus*, and others which have a positive inotropic effect on the heart are still the drugs of choice for congestive heart failure. The anticoagulants dicoumarol and other coumarin derivatives have been devel-oped from plants; the ergot alkaloid ergotamine is a vasoconstrictor used in the treatment of severe vascular headache; alkaloids from *Veratrum* and *Rau-wolfia* species are antihypertensive; and the tannins, which occur in many plant species, are haemostatic. Quinidine, from *Cinchona* bark, was the main antifibrillatory agent in use until the advent of modern anti-arrhythmics and is still used extensively. Classical screening tests for CV activity are well docu-mented; references to work on these are included.

TESTING FOR ANTIHYPERTENSIVE ACTIVITY

The mode of action of antihypertensive agents can be complex, since the cause of hypertension itself is often multifactorial: genetic disposition, diet, stress, age, drug and alcohol abuse, smoking, and other previous or co-existing diseases. There are a number of laboratory models used to detect blood pressure-lowering activity which can be used to test plant extracts with potential for use in the treatment of hypertension. Blood pressure is a

69

function of cardiac output and peripheral resistance; the latter is controlled in part by the calibre of resistance vessels. Many antihypertensive drugs in use act directly on blood vessel smooth muscles as vasodilators by a variety of mechanisms including, with examples:

1. *α_1-adrenergic block*: prazosin, phentolamine, phenoxybenzamine.
2. *Release of nitric oxide* in situ, *which by increasing the local concentration of cyclic-GMP causes hyperpolarisation of vascular smooth muscle*: glyceryl trinitrate, isosorbide mono- and dinitrates.
3. *Calcium channel blockade*: verapamil, nifedipine, diltiazem.

Other mechanisms of action of antihypertensives include:

4. *Inhibition of angiotensin converting enzymes (ACE-I and ACE-II)*: captopril, enalapril, losartan.
5. *β-blockers*: atenolol, propranolol (also used for arrhythmias).
6. *Centrally acting drugs*: methyldopa, clonidine.

If antihypertensive activity is found in the plant extract using experiments given, reference to 'Further Reading' should be made for methods of elucidating mechanism of action. This has been extended specially, because the field of cardiovascular pharmacology is so important and expanding rapidly.

Animal Models: Antihypertensives (Figures 6.1–6.7; Table 6.1)

Useful perfusion models are:

IN VITRO METHODS

Rabbit Ear Artery Preparation

The central artery of the severed ear is perfused with warm physiological saline solution (De la Lande and Rand, 1965; Day and Dixon, 1971).

Rat Mesenteric Artery

The perfusion system is connected to a suitable device for detecting changes in perfusion pressure to indicate vasoconstriction or vasodilation (MacGregor, 1965; Adeagbo and Okpako, 1980).

Rat or Rabbit Aorta

Isolated arterial preparations can be used to assess direct action on the circular muscle of the artery: a helical strip of the artery is mounted and contractions of the circular muscle are monitored by a strain gauge and recording device. Alternatively, an intact tube of the artery is mounted by means of stainless steel wires; this enables the contraction or relaxation of the circular muscle to be monitored. One advantage of this preparation is that it can be used with the endothelium intact or with the endothelium denuded. In this way, relaxation due to the release of endothelium-derived relaxing factor (EDRF) (Furchgott, 1983) or nitric oxide (Moncada *et al*, 1991) can be determined. An example of this method is illustrated below.

IN VIVO METHODS

Anaesthetised Normotensive Rat

The rat is anaesthetised with sodium pentabarbitone injected i.p. The trachea is exposed for artificial respiration and the carotid artery exposed and cannulated for arterial blood pressure measurement. One arm of the carotid artery cannula is connected to a pressure transducer and a pressure gauge for recording changes in blood pressure. Changes in heart rate in response to drugs can also be recorded at the same time. Drugs and vehicle are administered through the femoral or jugular vein. Tracings of an experiment are shown below.

Hypertensive Rats

Blood pressure (BP) should be above 160 mmHg (normal BP: 90–120 mmHg). It can be measured using a tail-cuff sphygmomanometer.

Spontaneously hypertensive rats (SHR) can be bought (Kiviranta *et al* (1989), Kamanyi *et al* (1993)), or hypertension can be induced by various methods, e.g.:
2-Kidney 1-clip hypertensive model. This is a surgically produced model where the activation of the renin–angiotensin system induces hypertension (*see* Victor, 1977; Yamahara *et al*, 1989);
DOCA/salt hypertensive model. The rat has a kidney removed under anaesthetic and is treated with deoxycortisone (DOCA) with salt until hypertension is induced (Yamahara *et al* (1989));
Salt-loaded hypertensive model. Normal food, and water containing 6% sodium chloride are given *ad libitum*. Hypertension takes about 10 days to appear (Kamanyi *et al* (1993)).

Treatment of Results

Antihypertensives

In vitro *Dilation of Rabbit Aorta Precontracted with Noradrenaline—Trace Obtained (Figure 6.1)*

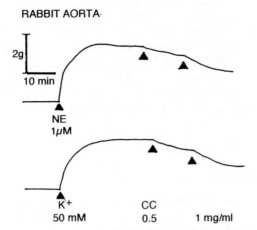

Figure 6.1 Representative tracings showing concentration-dependent inhibitory effects of *Capparis cartilaginea* (CC) on rabbit aorta precontracted with norepinephrine (NE), or K^+. From Hypotensive and spasmolytic activities of ethanolic extract of *Capparis cartilaginea*. Gilani, A.H. and Aftab, K. (1994). *Phytother. Res.*, **8**(2), 145–148, with permission.

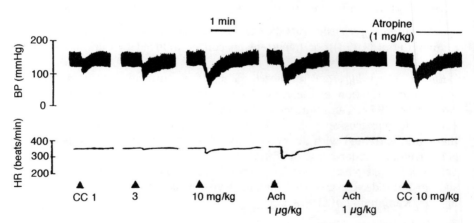

Figure 6.2 Comparison of *Capparis cartilaginea* (CC) and acetylcholine (Ach) for their effects in anaesthetised rats on blood pressure (BP) and heart rate (HR) in the absence and presence of atropine. Atropine was administered 5 min before the re-administration of Ach or CC. From Hypotensive and spasmolytic activities of ethanolic extract of *Capparis cartilaginea*. Gilani, A.H. and Aftab, K. (1994). *Phytother. Res.*, **8**(2), 145–148, with permission.

Table 6.1 Effect of ethanolic extract of *Capparis cartilaginea* on mean arterial blood pressure in normotensive anaesthetised rats

Dose (mg/kg)	Number of observations	% Fall in blood pressure ($\pm$ SEM)
1	5	20.3 ± 2.4
3	6	30.4 ± 3.3
10	6	47.2 ± 4.2

From Hypotensive and spasmolytic activities of ethanolic extract of *Capparis cartilaginea*. Gilani, A.H. and Aftab, K. (1994). *Phytother. Res.*, 8(2), 145–148, with permission.

Cardiovascular System: Blood Pressure

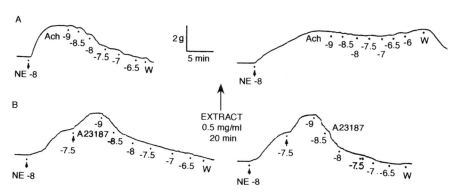

Figure 6.3 Representative original recordings showing the endothelium-dependent vasorelaxant action caused by acetylcholine (A) and calcium ionophore A23187 (B) in rabbit aortic rings precontracted by noradrenaline in the absence or presence of HE extract from *Hymenaea martiana* incubated in the bath for 20 min. Typical recording of at least five experiments. W indicates washing of the preparations. From Vascular action of the crude hydro-alcoholic extract (HE) from *Hymenaea martiana* on the isolated rat and rabbit aorta. Calixto, J.B. *et al* (1992), *Phytother. Res.*, 6(6), 327–331, with permission.

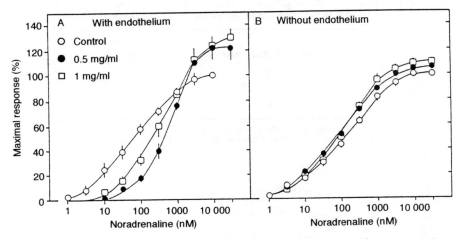

Figure 6.4 Effect of HE from *Hymenaea martiana* on mean contractile concentration–response curves induced by noradrenaline in rings of rabbit aorta with (A) or without (B) endothelium. Each point represents the mean of 5 to 6 experiments and the vertical bars the SEM. From Vascular action of the crude hydro-alcoholic extract (HE) from *Hymenaea martiana* on the isolated rat and rabbit aorta. Calixto, J.B. *et al* (1992), *Phytother. Res.*, **6**(6), 327–331, with permission.

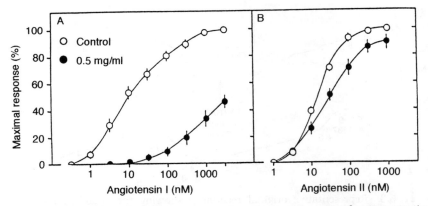

Figure 6.5 Effect of HE from *Hymenaea martiana* on mean contractile concentration–response curves induced by angiotensin I (A) and angiotensin II (B) in rings of rabbit aorta set up with intact endothelium. Each point represents the mean of 5 to 6 experiments and the vertical lines indicate the SEM. From Vascular action of the crude hydro-alcoholic extract (HE) from *Hymenaea martiana* on the isolated rat and rabbit aorta. Calixto, J.B. *et al* (1992), *Phytother. Res.*, **6**(6), 327–331, with permission.

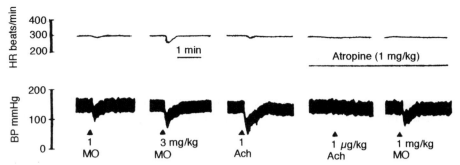

Figure 6.6 Tracing from a typical experiment showing the dose-dependent hypotensive effect of compound 1 of *Moringa oleifera* (MO) in anaesthetised rats. Effect of atropine pretreatment on MO and acetylcholine is also shown. Similar results were found with other compounds. From Pharmacological studies on hypotension and spasmolytic activities of pure compounds from *Moringa oleifera*. Gilani, A.H. *et al* (1994), *Phytother. Res.*, 8(2), 87–91, with permission.

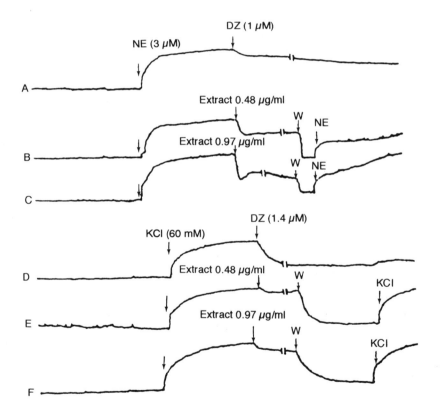

Figure 6.7 Effect of *T. arjuna* on the norepinephrine (3 μM) and KCl (60 mM) induced contraction of rat aorta. A–C, contraction induced by norepinephrine; D–F, contraction induced by KCl. From *Terminalia arjuna* extract modulates the contraction of rat aorta induced by KCl and norepinephrine. Tripathi, Y.B. (1993), *Phytother. Res.*, 7(4), 320–322, with permission.

TESTING FOR EFFECTS ON THE HEART: DIRECT ACTION ON THE MYOCARDIUM

Some drugs act directly on the heart. The sympathetic drive to the heart is mediated by noradrenaline (NA; norepinephrine) acting on β_1-adrenoceptors. The vagal control of the heart is mediated by acetylcholine (ACH) via muscarinic receptors. Substances interacting with these and other receptor systems, or with membrane cation gating systems such as calcium channels, or intracellular second messengers (cAMP), may stimulate or depress myocardial function directly. Substances that release catecholamines from adrenergic nerve endings, such as tyramine, will also stimulate the heart.

A number of *in vitro* methods are available for testing for a direct action of drugs on the myocardium. If a plant extract depresses the myocardium this may be indicative of an undesirable cardiotoxicity, but it may suggest potential therapeutic applications as an anti-arrhythmic or antihypertensive agent. A substance causing an increase in the rate and force of the contraction of the myocardium (positive chonotropic and inotropic effects) may be potentially useful as a cardiotonic, or this may be an indication that the drug may cause undesirable arrhythmias in clinical usage. The interpretation of the data in terms of potential therapeutic application of the plant extract must therefore depend on the total pharmacological profile of action of the extract. On the other hand, the *in vivo* methods described below can be used to evaluate the possible mechanism of action of the extract at the molecular level, even if its potential clinical application lies elsewhere.

Animal Models: Direct Action on the Myocardium

IN VITRO METHODS

Isolated Perfused Guinea-pig or Rabbit Heart (Langendorff Method) (Figure 6.8)

This preparation is set up as described by Burns (1952). The animal is killed by a blow to the head and cutting the neck vessels, but ensuring that a good length of the stump of the aorta is left. The heart is removed to a Petri dish containing PSS (Krebs bicarbonate) prewarmed to 37 °C and saturated with a mixture of 5% CO_2 in oxygen. The whole heart is perfused through a cannula inserted into the aorta. By means of a clip attached to the tip of the left ventricle, the heart is connected through pulleys to a force displacement transducer and recording device. A side arm of the perfusion cannula is connected to a pressure transducer to measure coronary circulation pressure changes. Drug solutions are administered directly into the aorta. The whole preparation should be enclosed in a constant temperature chamber. More details and applications of this preparation can be found in Friedrichs *et al*

(1994). The main advantages of the Langendorff preparation are that it measures the left ventricular rate and force of contraction and the effect of drug or extract on coronary flow and also on left ventricular pressure.

Rat or Guinea-pig Isolated Atria (Figures 6.9–6.11)

When isolated with the sino-atrial node (pacemaker) intact, the atria continue to beat for many hours when incubated in a suitable PSS and connected to a force displacement transducer and a recording device. Changes in the rate and force on contraction caused by drugs or plant extract added to the organ bath can be recorded. The two atria (left and right) from rat or guinea-pig can be set up as a paired preparation or, quite commonly, the right atrium only can be used. The PSS should be well aerated with 5% CO_2 in oxygen.

Tracings of an experiment are shown.

IN VIVO METHOD (Figure 6.12)

The effect of an extract on the cardiovascular system can also be measured using a whole animal. The rat, anaesthetised with i.p. sodium pentobarbitone, is usually used for this. The trachea is exposed for artificial respiration, and the carotid artery exposed and cannulated for arterial blood pressure measurement. One arm of the carotid artery cannula is connected to a pressure transducer, and a gauge for recording changes in blood pressure. Changes in heart rate in response to drugs can also be recorded at the same time. Drugs and drug vehicle are administered through the femoral or jugular vein.

Tracings of an experiment are shown below.

Arrhythmia can be induced by acetylcholine (Ach) and low K^+ levels in isolated atria of guinea-pig or rabbit, arranged so they can be stimulated by high frequency. Fibrillation is induced by exposure to 300 µg/ml Ach and low K^+ (1/8 normal) and stimulated at 1200 pulses/min. The atria go into fibrillation which is prevented by quinidine (McLeod and Reynolds, 1962).

Anti-arrhythmic Drugs

The term arrhythmia is used to describe disorders of heart rate or rhythm. These are usually caused by impulses originating from ectopic foci or by abnormal conduction of normally generated impulses. Drugs used to restore rhythm and rate are known as anti-arrhythmic (or antidysrhythmic) drugs and their pharmacological properties are complex. There are various ways of grouping these drugs, but the classification suggested by Vaughan Williams (1975) remains widely accepted. This classification is based on the action of the drug on various functional properties such as *automaticity, conductivity,*

Table 6.2   Classification of commonly used anti-arrhythmic drugs

Class	Mechanism of action	Drug examples
I	Membrane stabilisation (local anaesthetic action)	Mexiletine, quinidine, procainamide, lignocaine
II	Reduced adrenergic influence on the heart, β-blockade, or adrenergic neurone blockade	Atenolol, timolol, bretylium
III	Prolongation of cardiac action potential	Amiodarone
IV	Calcium channel blockade	Verapamil, nifedipine
V	Increased vagal tone	Digitalis, edrophonium, methoxamine

and *repolarisation* characteristics of myocardial tissues. This is regardless of any other properties the drug may have. Using these criteria, the commonly used anti-arrhythmic drugs fall into five classes as in Table 6.2.

This classification is by no means rigid, as some Class I drugs (e.g. quinidine) also have Class III properties, and so on. Also the clinical uses of many anti-arrhythmic drugs are not restricted to cardiac arrhythmias. Many of the drugs named in the table are also antihypertensives, vasodilators. Others such as those in Class V are anticholinesterase, vasoconstrictors and cardiac glycosides.

What this means is that it is not a simple matter to design a method specifically to screen for anti-arrhythmic properties in a plant extract. Some inference as to anti-arrhythmic potential can be made from the various tests already described for action on the cardiovascular system, and then using references listed in 'Further Reading'.

Treatment of Results: Direct Action on the Myocardium

Isolated Perfused Heart (Langendorff Preparation) Of Rabbit in vitro. Tracing Obtained (Figure 6.8).

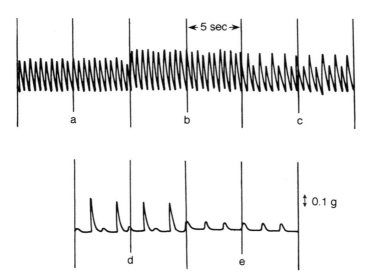

Figure 6.8 Typical recordings of the effect of aqueous extract of *Baphia nitida* on the isolated rabbit heart preparation. (a) Control, (b) 5.0×10^{-4}, (c) 1.0×10^{-3}, (d) 2.0×10^{-3}, and (e) 4.0×10^{-3} g/ml, respectively, of the extract. Each extract concentration was allowed to act for 3 min. From Effects of aqueous extract of *Baphia nitida* on isolated cardiac tissues. Adeyemi, O.O. (1992), *Phytother. Res.*, 6(6), 318–321, with permission.

Isolated Atria of Rat. Tracing Obtained (Figure 6.9) and Dose–Response Curve to $CaCl_2$ of Plant Extract (Figure 6.10).

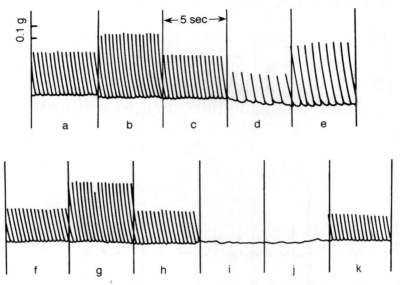

Figure 6.9   Effect of extract on the atrial responses to calcium chloride ($CaCl_2$). (a) Control, (b) 10 mM $CaCl_2$, (c) control, (d) 2.5×10^{-2} g/ml extract, (e) 10 mM $CaCl_2$ in the presence of 2.5×10^{-2} g/ml extract, (f) control, (g) 10 mM $CaCl_2$, (h) control, (i) 5.0×10^{-2} g/ml extract, (j) 10 mM $CaCl_2$ in the presence of 5.0×10^{-2} g/ml extract and (k) control after washing off extract and $CaCl_2$. From Effects of aqueous extract of *Baphia nitida* on isolated cardiac tissues. Adeyemi, O.O. (1992), *Phytother. Res.*, **6**(6), 318–321, with permission.

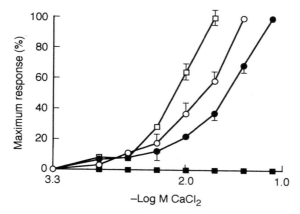

Figure 6.10 Log concentration–response curves to $CaCl_2$ in the absence □—□, and presence of 5.0×10^{-3} g/ml ○—○, 2.5×10^{-2} g/ml ●—● and 5.0×10^{-2} g/ml ■—■ of *Baphia nitida* extract on the isolated rat atria preparation. Each extract concentration was left in contact with the tissue for 15 min before $CaCl_2$ additions. Values represent mean ± SEM ($n = 6$). From Effects of aqueous extract of *Baphia nitida* on isolated cardiac tissues. Adeyemi, O.O. (1992), *Phytother. Res.*, 6(6), 318–321, with permission.

In vitro: *Isolated Spontaneous Beating Paired Atria from Guinea-pig*: Tracings Obtained (Figure 6.11)

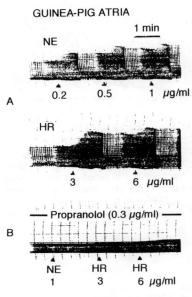

Figure 6.11 A representative tracing showing comparison of norepinephrine (NE) and the plant extract (HR) in the absence (A) and presence (B) of propranolol in isolated spontaneous beating paired atria from guinea-pigs. Propranolol was administered 10 min before the redetermination of NE or HR responses. From Vasoconstrictor and cardiotonic actions of *Haloxylon recurrum* extract. Gilani, A.H. and Shaheen, F. (1994), *Phytother. Res.*, **8**(2), 115–117, with permission.

In vivo *Method: Anaesthetised rat.* Tracings Obtained (Figure 6.12)

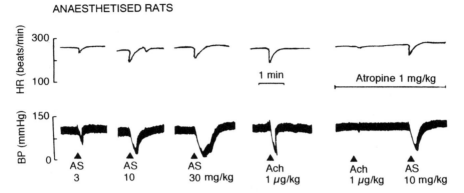

Figure 6.12 Comparison of *Artemisia scoparia* (AS) and acetylcholine (ACh) for their effects on blood pressure (BP) and heart rate (HR) in the absence and presence of atropine in anaesthetised rats. Atropine was administered 5 min before the administration of ACh or AS. From Ca^{++} channel blocking activity of *Artemisia scoparia* extract. Gilani, A.H. *et al* (1994), *Phytother. Res.*, **8**(3), 161–165, with permission.

ANTIPLATELET AND ANTITHROMBOTIC ACTIVITY

Platelet aggregation studies are a useful and convenient method of investigating not only antiplatelet and antithrombotic mechanisms, but also those involved in other conditions such as inflammation and migraine.

THROMBOSIS

Thrombosis (blood coagulation) is a complex event, part of which involves platelet activation and aggregation. Platelets (also called thrombocytes) may be activated on exposure to hormones such as adrenaline and vasopressin; autacoids including ADP (adenosine diphosphate); serotonin (5-hydroxytryptamine, 5-HT); eicosanoids; PAF (platelet-activating factor); blood coagulation and complement factors (thrombin and plasmin); and vascular proteins (collagen, elastin and others). These stimuli interact with receptor sites on the platelet and activate intracellular metabolic pathways. Second messenger systems involving cyclases, protein kinase C and calcium channels will then result in liberation of arachidonic acid from membrane phospholipids and ultimately cause platelet aggregation and thrombus formation.

There are several *in vivo* systems for the investigation of thrombosis; however, most of these are specialised and beyond the scope of this book (*see* 'Further Reading'). Some useful *in vitro* models are described below.

PLATELET AGGREGATION

When platelets interact with substances such as collagen, ADP, thrombin, PAF (another important inflammation mediator which will be discussed separately), or when they adhere to a damaged vessel wall, aggregation may occur. This is at first reversible but may then lead to a second, irreversible aggregation, during which pharmacologically active substances are released.

These substances include: *ADP*, which induces other platelets to aggregate and release more ADP; *5-HT (5-hydroxytryptamine or serotonin)*, which induces further aggregation; *Prostanoids* (prostaglandins and related compounds), especially thromboxane A_2 (TXA_2), which is produced by thromboxane synthase from the products of cyclo-oxygenase-induced metabolism of arachidonic acid as in the chart.

The main functional responses of platelets are adhesion, aggregation and secretion. Adhesion occurs during exposure of the platelet to a thrombogenic surface (i.e. 'wetting'). The platelet then spreads. If the stimulus is soluble, e.g. ADP, thrombin, TXA_2 and PAF, the platelet will undergo a shape change from discoid to spherical, extend pseudopods and aggregate. So far the process is reversible and the platelet intact. Secretion may then occur, where platelets release the contents of their granules. This is termed the *platelet release reaction* in which thrombospondin, fibrinogen, fibronectin and the aggregating agents ADP and serotonin lead to further irreversible aggregation and the build-up of the thrombus.

METABOLISM OF ARACHIDONIC ACID (Figure 6.13)

The metabolism of arachidonic acid is crucial to platelet activation as well as being an intrinsic part of the inflammatory response. Most anti-inflammatory drugs are taken for disorders such as arthritis and migraine: therefore anti-inflammatory testing will be discussed in a separate section. Platelet function can, however, be studied to partly elucidate these mechanisms of action since they have common metabolic pathways and fairly specific inhibitors are available to block each stage involved.

Figure 6.13 shows the major arachidonate metabolic pathways; these are common to other inflammatory systems, and their understanding is relevant when considering the wider physiological implications. It is not always simple: e.g. aspirin is effective at low doses as an antithrombotic agent by inhibiting cyclo-oxygenase in platelets but not in vascular endothelium. The explanation for this difference is not fully known but it is important because prostacyclin synthase in vascular endothelium acts upon a cyclo-oxygenase product, PGH_2, to form prostacyclin (PGI_2), which is a potent antithrombotic agent (*see* Figure 6.13 for clarification). Low dose aspirin can thus prevent thrombosis without interfering with PGI_2-induced vasodilation.

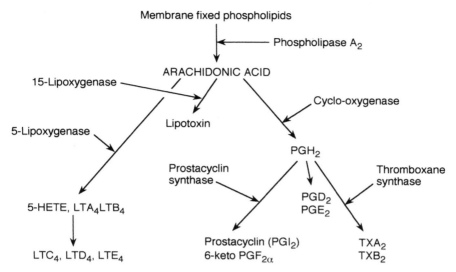

Membrane fixed phospholipids

Phospholipase A_2

ARACHIDONIC ACID

15-Lipoxygenase

Cyclo-oxygenase

5-Lipoxygenase

Lipotoxin

PGH_2

Prostacyclin synthase

Thromboxane synthase

5-HETE, LTA_4LTB_4

PGD_2
PGE_2

LTC_4, LTD_4, LTE_4

Prostacyclin (PGI_2)
6-keto $PGF_{2\alpha}$

TXA_2
TXB_2

Figure 6.13   Arachidonic acid metabolism. TX = thromboxane; PG = prostaglandin; LT = Leukotriene; HETE = hydroxyeicosatetraenoic acid.

Platelet Aggregation Models

In vitro platelet preparations are an example of a whole cell system providing information under physiological conditions, and as such hold advantages over isolated enzyme systems such as cyclo-oxygenase and lipoxygenase. Inhibitors of these enzymes can be used in platelet aggregation experiments.

As well as being useful as screens for anti-aggregating agents, platelet function can be used to investigate the mechanism of action of those compounds causing aggregation and inflammation, such as the phorbol esters, which are important biochemical tools as well as environmental hazards. Serotonin, released by platelets, is involved in the aetiology of migraine, and the system has been used to elucidate the mechanism of action of feverfew (*Tanacetum parthenium*), a plant preparation used as a migraine prophylactic (*see* examples of both in the 'Results' section).

The usual method of preparing platelets is by centrifuging whole blood (with an added anticoagulant such as heparin or citrate) to produce *platelet-rich plasma* (PRP). This can be used as it is, or the platelets can be further centrifuged and resuspended in a special medium to prepare *washed platelets*. The advantage of using PRP is that it is quick and involves minimum manipulation (which can lead to activation) of the platelets; washed platelets, however, are stable for longer and are free of anticoagulant and plasma proteins.

The study of aggregation uses a turbidimetric method: as platelets clump together the transmission of light increases through the suspension and is measured by a photometer. An ordinary photometer would not be suitable because the platelets must be stirred during the experiment and kept at 37 °C. The *platelet aggregometer* has been developed to do this and is commercially available. Basic ones are adequate but multichannel aggregometers enable more experiments to be done in the time available and faster time–action curves constructed.

RECEPTOR STUDIES USING INTACT PLATELETS

Receptor binding on platelet membranes is the usual mechanism by which platelet activation occurs; these include the α_2-adrenoceptor and the PAF receptor. It is not within our scope here to go into radioreceptor assays and the production of membrane preparations, and in fact many authors recommend intact platelets in an aggregometer as the method of choice. The following receptor studies also illustrate the use of platelet aggregation methods for investigating effects unrelated to antithrombotic activity.

α-ADRENOCEPTOR

A method of screening has been developed to identify α-adrenergic antagonism by measuring the interference of a compound with the potentiation by adrenaline of ADP-induced platelet aggregation (Beretz and Cazenave, 1991). Such compounds include the classical α-adrenergic blockers yohimbine, α-yohimbine (rauwolscine) and dihydroergocryptine.

Platelet Activating Factor (PAF)

PAF is an ether phospholipid of several molecular species varying in the length of the *O*-alkyl side chain. It is in itself a weak inducer of platelet aggregation but it activates platelets induced by other agonists. It is important in the aetiology of asthma, allergic inflammation and anaphylactic shock. It binds to a specific receptor on the platelet surface and the binding affinity studies appear to correlate well with those involving platelet aggregation.

At present the main PAF-inhibitors are natural products, such as the ginkgolides from *Ginkgo biloba* and kadsurenone from *Piper futokadsurae*. For investigation of these, standard platelet techniques can be used but it may be necessary in some cases to inhibit the arachidonate pathway (by adding aspirin) and secreted ADP (by adding apyrase). It may also be useful in this instance to use rabbit platelets for PAF work since they are at least 10 times more sensitive to PAF than human platelets. For more about PAF-antagonists *see* Chapter 7, 'The Respiratory System'.

Materials and Methods: Platelet Aggregation (Figures 6.14–6.16; Table 6.3)

Reagents: Platelet Aggregating Agonists and Antagonists

Reagents are all readily available, e.g. from Sigma, Aldrich, pharmaceutical manufacturers, etc., unless specified. Suitable solvents are normal saline, Tyrode buffer (*see* Appendix II) ethanol and acetone. If in doubt check solvents on platelets before experiment. Chloroform and surfactants must not be used as they disrupt the platelet membrane.

Platelet Aggregating Agonists: Preparation, Dose Range, Mechanisms of Action and Use

ADP   Causes platelet shape change leading to aggregation. Concentration range (in platelets), $0.1–10.0\,\mu M$ in Tyrode buffer or ethanol. Store at $-30\,°C$.

Adrenaline   Adrenergic agonist. Concentration range, $1–100\,\mu M$.

Arachidonic acid   Metabolised to prostaglandin endoperoxides and thromboxane A_2 by cyclo-oxygenase, causing release reaction. Concentration range, $0.3–5.0\,mM$, in $Na_2CO_3\,mM$ or ethanol. Store at $-30\,°C$.

A23187   Calcium ionophore available from Boehringer. Stock solutions in DMSO stored at $-30\,°C$, diluted when required with saline (0.9% NaCl). Concentration range, $10–100\,\mu M$.

Collagen   Activates platelet release reaction. Suspended in saline. Concentration range, $1–10\,\mu g\;ml^{-1}$.

PAF (platelet activating factor)   Acts at PAF receptor. Available from Bachem, Switzerland. Concentration range $1–10\,\mu M$. Stock solutions in 80/20 chloroform/methanol stored at $-4\,°C$, evaporated down and resuspended in Tyrode albumin buffer for use, stored at $-30\,°C$. Remove all chloroform before adding to platelets.

Phorbol esters   TPA (tetradecanoyl phorbol acetate, formerly called PMA, phorbol myristate acetate) available commercially. Others (e.g. 12-DOPP, 12-deoxyphorbol phenylacetate) isolated from latex of *Euphorbia* spp. Activates protein kinase C. Concentration TPA in nM range; 12-DOPP $1–20\,\mu M$.

Thrombin   Receptor binding activates platelet release reaction. Suspended in saline. Concentration, $1–10\,Uml^{-1}$.

U46619   Thromboxane mimetic. Concentration range, $3–10\,\mu M$.

Antiplatelet Aggregating Agents: Preparation, Dose Range, Mechanism of Action and Range   This list comprises a selection of reagents which can be used to elucidate the mechanism of action of a substance causing platelet aggregation. The concentration range is only a guide and depends to some extent on the aggregating agent used.

Aminopyrine   Free radical (FR) scavenger; prevents cyclo-oxygenase involvement since FR formation is an initiating step in prostaglandin synthesis. Concentration range, 0.2–2.0 mM.

Aspirin   Cyclo-oxygenase inhibitor; prevents arachidonic acid-induced aggregation. Used in PAF experiments to make sure cyclo-oxygenase pathway is not involved. Concentration range, 1–50 mM.

BN 52021 (Ginkgolide B); BN 52063 (Ginkgolide mixture)   PAF-antagonist; ginkgolide B being the most potent. Isolated from the maidenhair tree, *Ginkgo biloba*. Available from Ipsen International Ltd and others.

Colchicine   Inhibits microtubule assembly of platelets. Prevents phorbol ester-induced aggregation, unlike many other antagonists. Concentration range, 1 μM–1 mM.

Clotrimazole   Thromboxane synthetase inhibitor. Concentration range, 0.1–1.0 mM.

EDTA   Chelates divalent cations (calcium in this case). Range, 3–5 mM.

Indomethacin   Cyclo-oxygenase inhibitor. Concentration range, 0.03–1.0 mM.

Pinane thromboxane A_2 and carbocyclic thromboxane A_2   Thromboxane analogues; thromboxane and prostaglandin endoperoxide receptor antagonist. Concentration range, 1–4 μM. Not readily available.

Phenidone   Cyclo-oxygenase/lipoxygenase inhibitor. Concentration range, 0.1–1.0 mM.

Propranolol   Phospholipase A_2 inhibitor. Prevents release of arachidonic acid from platelet membranes and therefore aggregation via cyclo-oxygenase pathway. Concentration range, 0.05–5.0 mM.

Prostacyclin (PGI_2)   Elevates cyclic AMP levels, preventing platelet aggregation by most agonists, including phorbol esters. Available from Upjohn. Concentration range, 1–10 nM. Store at −30 °C.

Trifluoperazine   Calmodulin inhibitor. Inhibits aggregation induced by most agonists. Concentration range, 0.05–0.5 mM.

Verapamil   Calcium antagonist. Prevents aggregation by most agonists, but not phorbol esters. Concentration range, 0.003–0.3 mM.

Yohimbine   α-Adrenergic antagonist. Prevents adrenaline-induced aggregation. Concentration range, 0.2–0.5 mM. For use in receptor binding studies on the α-adrenoceptor, *see* text.

Preparation of Platelets

Human Platelet-rich Plasma (PRP)   Blood is obtained from the forearm vein of donors (usually male) who have not taken any drugs, e.g. aspirin, likely to interfere with platelet function. Blood must be taken carefully, using a large diameter needle (18/10), to avoid damaging platelets during venipuncture.

Blood is collected into an anticoagulant such as trisodium citrate, which lowers Ca^{2+} levels and prevents thrombin generation. The final concentration should be 0.38% (13 mM), or 9:1, blood:trisodium citrate 3.24%. Heparin is not suitable as it can potentiate platelet agonists, but natural hirudin from the medicinal leech *Hirudis medicinalis* or recombinant hirudins are a possibility. The tube is gently inverted to mix, and the citrated blood centrifuged at $160–175 \times g$ for 10–20 min to obtain PRP. The supernatant, a layer of translucent straw-coloured serum, is the PRP. This is removed and kept for at least 30 min to stabilise while the remaining lower layer containing the erythrocytes, etc. is recentrifuged at a much higher speed, e.g. $10\,000 \times g$ for 2 min, or $2700 \times g$ for 20 min, until a transparent layer of platelet-poor plasma (PPP) is obtained. This is removed with a pipette into the cuvette of the aggregometer to act as the 'blank' for light transmission measurements. Many workers then adjust the platelet count of the PRP (with PPP) to 300 000 platelets μl^{-1}, but in fact this standardisation is not necessary when comparing results performed on the same sample of blood. Platelets must be used within about 3 h to guarantee reproducible responses, but it is a simple matter to do a few aggregations with ADP or arachidonic acid, for example, to ensure that the platelets are still performing as expected. Further details are given in the example shown.

Rabbit Platelets These may be used for a number of reasons such as ethical/legal, health (HIV/hepatitis) considerations, and for comparative purposes. Blood is collected from the marginal ear vein by needle and syringe or as follows: Clip the hair from below the point on the vein where the cut is to be made. Apply some petroleum jelly to the ear (to stop the blood 'wetting' the surface and coagulating) and dab a little xylene on cotton wool onto the vein to bring it up. Then nick the vein with a scalpel and allow the blood to flow into the citrate-containing vessel, all quantities as before.

About 20 ml of blood should be taken (this will give at least 20 aggregations and not harm the rabbit, which can be used again). Then clamp the blood vessel between the finger and thumb with a small pad of cotton wool dampened with ethanol and hold until bleeding stops. PRP and PPP are prepared in the same way as before.

Washed Platelets This involves more steps than for PRP, and requires the addition of reagents such as PGI_2, heparin, apyrase and others at various stages to prevent platelet activation. All steps are carried out at room temperature. Because of the extra manipulations, it is only worth doing when it is necessary to remove all traces of plasma proteins (which may bind compounds) or run experiments for a longer duration than is possible for PRP. If desired, the details of the method are available (Beretz and Cazenave, 1991).

Preparation of Plant Extracts

An extract must be solubilised and, if necessary, either centrifuged or filtered before adding to the PRP in the aggregometer. Aqueous plant extracts usually contain a high amount of calcium ions which will initiate aggregation, and chloroform and some other organic solvents will disrupt the platelet membrane; such extracts should be evaporated down and solubilised in a suitable solvent, e.g. ethanol, acetone, DMSO, hexane. Too much 'ballast', e.g. tannins, pigments, etc., will also interfere and need to be removed or taken into account by using a control.

Testing the Extracts: Platelet Aggregation

The manufacturer's instructions for using the aggregometer should be followed, testing the system by performing several aggregations using standard agonists such as ADP, collagen, thrombin and arachidonic acid. The extract should then be added to the platelets alone, to check that it does not itself cause aggregation, and if in doubt, to the solvent alone. To test for anti-aggregating activity, the PRP is pretreated with the extract about 1 min before a standard aggregating agent is added. The dose of a crude extract employed is arbitrary and will need adjusting in the light of preliminary results. Aggregations are normally followed for up to 5 min.

If aggregation occurs with the plant extract alone, then standard antagonists (prostacyclin, etc.; *see* list above) can be used to give some idea of which systems are being activated.

Measurement of Platelet Aggregation

It can be seen from the traces of aggregation curves shown that light transmission increases as platelets aggregate. The maximum height of the curve is taken as 100% platelet aggregation after a specified time interval (usually 3 min) and any reduction of this due to the presence of the drug expressed as a percentage inhibition. The percentage inhibition can be plotted against log concentration to obtain a sigmoid dose–response curve in the usual way, and the IC_{50} (concentration causing 50% inhibition) deduced for comparative purposes.

Treatment of Results

Inhibition of Platelet Aggregation: Tracings Obtained (Figure 6.14) and Expression of Results (Table 6.2)

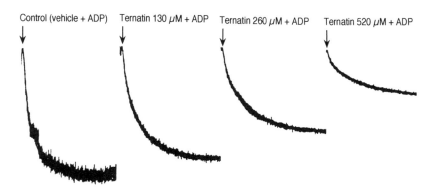

Figure 6.14 Typical tracing showing the effect of ternatin on ADP (10 μM)-induced aggregation to guinea-pig PRP. From Antithrombotic activity of ternatin, a tetramethoxy flavone from *Egletes viscosa*. Souza, M.F. *et al* (1994), *Phytother. Res.*, 8(8), 478–481, with permission.

Experimental Details and Explanation of Tracing

Platelet rich plasma (PRP) was obtained from blood collected from guinea-pig by venepuncture, anticoagulated with sodium citrate (3.8%, 9:1) and centrifuged at $150 \times g$ for 20 min. 0.45 ml aliquots of PRP were tested using ADP as an aggregating agent. ADP was dissolved in saline (154 mM NaCl) whereas the test compound ternatin was dissolved in DMSO (2% in 154 mM NaCl). DMSO at this concentration was shown not to interfere with the test. The concentrations of both ADP and ternatin were prepared at a concentration that gave the desired final plasma concentration when added to 0.45 ml of PRP. For ADP, this corresponds to 10 μM and for ternatin 130, 260 and 520 μM. The PRP-test compound mixture was incubated for 5 min at 37 °C before adding ADP. Ternatin was found to inhibit ADP-induced platelet aggregation in a concentration-dependent manner (Table 6.3 and Figure 6.14). A significant difference from controls was present at concentrations of 200 and 400 μM, when the maximum degree of aggregation (i.e. the maximum transmission change) was taken into consideration. The calculated IC_{50} for ternatin was 390 μM.

Table 6.3 Effect of ternatin on *in vitro* ADP (10 μM)-induced platelet aggregation to guinea-pig PRP

Group	Concentration (μM)	Maximum transmission change (mm)	Inhibition (%)
Control	–	124.5 ± 8.3	–
Ternatin	130	98.0 ± 5.8	15
	260	76.8 ± 7.1[a]	40
	520	41.0 ± 6.0[a]	63

Significantly different from the control at [a]$p < 0.01$ (Student's *t*-test). From Antithrombotic activity of ternatin, a tetramethoxy flavone from *Egletes viscosa*. Souza, M.F. *et al* (1994), *Phytother. Res.*, 8(8), 478–481, with permission.

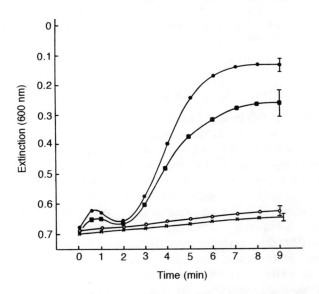

Figure 6.15 *In vitro* effect of 40 μl (■) and 60 μl (○) aqueous extracts of *Herba galegae* on platelet aggregation of 400 μl PRP by thrombin. The effect of 20 μl (15 units/ml) heparin (×) was used for comparison. For control (●) the aggregation of 400 μl PRP was used, after addition of 50 μl thrombin (8 units/ml). Values are means of ± maximum standard errors for 10 independent experiments. From Effect of the water extract of *Galega officinalis* on human platelet aggregation *in vitro*. Atanasov, A.T. (1994), *Phytother. Res.*, 8(5), 314–316, with permission.

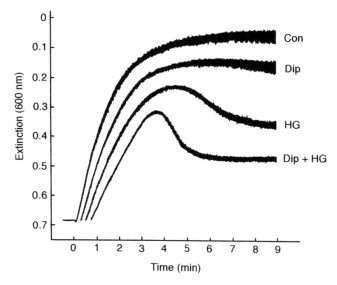

Figure 6.16 *In vitro* effect of separate application of 20 μl (5 mg/ml) dipyridamole (Dip) and 20 μl aqueous extract of *Herba galegae* (HG), and combined application of 20 μl dipyridamole and 20 μl extract (Dip + HG) on platelet aggregation of 400 μl PRP, after addition of 20 μl ADP (1 × 10⁻³ M). From Effect of the water extract of *Galega officinalis* on human platelet aggregation *in vitro*. Atanasov, A.T. (1994), *Phytother. Res.*, **8**(5), 314–316, with permission.

ANTISCLEROTICS: CHOLESTEROL AND LIPID-LOWERING ACTIVITY

It is now widely recognised that high serum levels of cholesterol and trigly-cerides, collectively known as hyperlipidaemia, is one of the main risk factors in the development of coronary heart disease (CHD) which is a product of atherosclerosis and thrombosis. Atherosclerosis is a long-term condition in which fatty deposits accumulate in the lining of arteries, causing narrowing of the lumen. Narrowing of the coronary artery may result in ischaemic pain, known as angina pectoris, during exercise. If a thrombus occurs where a blood vessel has already been narrowed by atherosclerosis it may block, and if this occurs in the coronary artery the result is a heart attack, or myo-cardial infarction. The development of a thrombus is outlined in the chapter on antithrombotics.

The role of diet in this process is of importance. Current dietary advice suggests that a low fat diet is desirable for prevention of CHD. It is compli-cated by the fact that hyperlipidaemia is also caused by an inherited defect in lipid metabolism and by disease, in which case drug treatment may be needed as well. There is still a great deal of debate about the effects of the different types of fat ingested, as well as the quantity, and this will be sum-marised briefly.

Saturated fat is generally agreed to be the most damaging and a low fat diet will lower blood cholesterol better than a low cholesterol diet. Dietary cholesterol has little influence on serum cholesterol in most people. Polyun-saturated fatty acids of the n-6 series have enjoyed a reputation as a benefi-cial substitute for saturated fats, but the synthetic *trans* fatty acids formed during the partial hydrogenation of oils to produce margarine are now thought to contribute to CHD by increasing the ratio of LDL to HDL (*see* Table 6.4), and high dietary levels are now not recommended. Polyunsatu-rates of the n-3 series, found in fish oils, are known to lower elevated serum triglyceride levels, protecting against CHD, but with no effect on serum cholesterol. They also have anticlotting activity. Mono-unsaturated fatty acids, such as those found in olive oil, are associated with low rates of CHD where they are the main source of fat (the 'Mediterranean diet') but

Table 6.4   Serum lipoproteins: content and function

Lipoprotein complex	Lipid content	Function occurrence	Features/significance of elevation
Chylomicron	ChL TG	Carries dietary TG and ChL to peripheral tissues	TG increased, ChL normal or slightly raised. Caused by lipoprotein lipase deficiency
VLDL (very low density lipoprotein)	TG	Carries TG from liver to periphery. Most serum TG in this form	TG increased, ChL slightly raised. Associated with CHD
IDL (intermediate density lipoprotein)	ChL	Breakdown product of VLDL	Transient particles. Some converted to LDL, some taken up by liver
LDL (low density lipoprotein)	ChL	Derived from VLDL. Carries ChL from liver to peripheral tissues	ChL increased, normal TG. Caused by abnormal or decreased LDL receptors. Tends to deposit ChL in artery walls. Associated with CHD
HDL	ChL	Carries excess ChL from peripheral tissues to liver for disposal	Elevated levels thought to protect against CHD

ChL = cholesterol; TG = triglyceride; CHD = coronary heart disease.

these diets usually involve other beneficial factors such as increased fruit and vegetable consumption. Then there are factors such as alcohol, smoking and complex carbohydrate intake which may contribute to CHD.

At present, plant products are more important in the dietary aspects of controlling hyperlipidaemia rather than as drug treatments, but garlic (*Allium sativum*), for instance, could be described as both. Other food substances which lower serum cholesterol are the seeds of fenugreek, *Trigonella foenum-graecum*, which are used as a condiment. Naturally occurring hypolipidaemics include the chromone khellin, from *Ammi visnaga*; the stilbene glucoside piceid, from *Polygonum cuspidatum*; the steroidal ketone, Z-guggulsterone, from *Commiphora mukul*; and various plant indoles.

Complex carbohydrate such as oat bran, guar gum and carob gums also lower serum cholesterol, as shown by several clinical studies, but their effects are being questioned since diets containing these are usually also low in fat. They may work by inhibiting fat absorption rather than by a direct pharmacological effect.

Lipid Metabolism and Disorders

Cholesterol is an essential component of cell membranes and a metabolic intermediate in the synthesis of steroid hormones and bile salts. Some cholesterol is obtained from the diet (about 25%) and the rest made by the liver, as required. High levels in serum may be a result of high dietary intake; this is known as exogenous hyperlipidaemia. Treatment of this type is by low fat diet and substitution of saturated fats as already described.

Endogenous hyperlipidaemia, a disorder of cholesterol metabolism, may be caused by genetic factors. This is primary or familial hyperlipidaemia and is much less common. Diseases such as diabetes mellitus, gout, hypothyroidism, obstructive jaundice, cirrhosis of the liver and renal failure can cause secondary hyperlipidaemia.

Fat from the diet is absorbed into the bloodstream and transported in the form of lipoproteins. These are classified into types (*see* Table 6.4) according to their density and their levels in serum used to classify primary hyperlipidaemias (or hyperlipoproteinaemias). This classification of disease is still evolving; *see* references cited in the General section of the 'Further Reading' list.

Elevated serum cholesterol *per se* is not the only indicator of atherosclerosis, since cholesterol levels may remain fairly normal while triglyceride levels rise in certain disorders, as shown in Table 6.4. It would therefore be preferable always to measure the levels of each lipoprotein fraction, especially LDL, all of which contain cholesterol. These were formerly specialist techniques but are increasingly being carried out now that kits are available. As shown in the table, HDL-cholesterol is thought to protect against CHD by 'mopping up' excess cholesterol. It is inversely related to total body cholesterol. The ratio of this to the lower density fractions is sometimes called the atherogenic index and is calculated as:

$$AI = \frac{\text{VLDL-cholesterol} + \text{LDL-cholesterol}}{\text{HDL-cholesterol}}$$

TESTING FOR ANTISCLEROTIC ACTIVITY: REDUCING SERUM LIPID LEVELS

Mechanisms of drug action in primary and secondary hyperlipidaemia may involve inhibition of cholesterol synthesis in the liver, e.g. by inhibiting hydroxymethylglutaryl-co-enzyme A (HMG-CoA) reductase, the rate-controlling enzyme in cholesterol biosynthesis. Several drugs acting through this mechanism are now in wide clinical use (e.g. simvastatin). Some compounds, such as bezafibrate, suppress endogenous cholesterol and triglyceride synthesis and stimulate catabolism via systemic lipoprotein and hepatic lipases; and the n-3 polyunsaturated fatty acids found in fish oils act by increasing lipoprotein lipase activity as well as inhibiting VLDL synthesis in the liver.

Some naturally occurring antioxidants such as flavonoids, with free-radical scavenging ability, have cholesterol-lowering activity in experimental hyperlipidaemia. The mechanism of action is thought to be at least partly due to inhibition of lipid peroxidation, but other factors affecting cholesterol control mechanisms must also be involved.

Cholesterol intestinal absorption can be inhibited using bile acid sequestrants, such as cholestyramine, which prevent the reabsorption of bile acids from the digestive system. This promotes hepatic conversion of cholesterol into bile acids and further breakdown of LDL-cholesterol. It also promotes faecal loss of dietary cholesterol. Unfortunately, these resins also interfere with drug and nutrient absorption, and patient compliance is poor due to the disgusting taste.

The mechanism of action of many hypolipidaemics is unknown, and therefore testing new plant products generally involves producing atherosclerosis experimentally and observing the changes, or measuring the effect on serum blood levels of cholesterol.

Animal Models: Lipid-lowering Agents

Atherosclerosis is induced with a high-fat, high-cholesterol diet over a period of 3–6 months, after which serum analysis can be carried out. *Hyperlipidaemia* occurs more rapidly and experiments take place over weeks rather than months. High sugar diets are sometimes used to induce atherosclerosis, especially if other metabolic disorders (e.g. diabetes mellitus) are under simultaneous investigation.

Methods given are for measuring blood lipid-lowering effects of plant extracts; related studies will be discussed and referenced.

Rabbits are frequently used as they are susceptible to atherosclerosis and have a similar bile acid metabolism to man. Hyperlipidaemia in rats may be induced by Triton administration, which interferes with uptake of plasma lipids.

Blood measurements taken may include total lipid, triglyceride, total cholesterol and lipoprotein cholesterol (especially LDL-cholesterol and HDL-cholesterol) and phospholipids. The latter is used because a high cholesterol: phospholipid ratio is associated with atherosclerosis. Faeces may also be collected and analysed to give an indication of fat and cholesterol excretion and possibly absorption.

Clinical studies using human volunteers normally involve ingestion of iso-calorific test and control diets over a period of time. These are ideal when the substance under test is an item of food or known to be innocuous.

Materials and Methods: Induced Hyperlipidaemia and Measuring Lipid-lowering Agents (Figures 6.17–6.20; Tables 6.5–6.8)

Animals

Rabbits, rats, in groups of > 6.

Atherogenic Diets

Examples of High Fat/Cholesterol Supplements to Normal Diet   May be administered separately by intragastric tube once daily, or incorporated into diet pellets.

1.   5% Cholesterol, 0.5% cholic acid in arachis oil, dose: 5 ml/kg.
2.   400 mg/kg Cholesterol in 5 ml coconut oil.
3.   2% Cholesterol, 0.5% cholic acid, 20% sunflower oil, expressed as a percentage addition to the normal diet.
4.   2% Cholesterol, 0.5% cholic acid as a percentage of normal diet.

Methods

Diet-induced Hyperlipidaemia and Blood Collection   Animals are fed the atherogenic diet for the required period (as little as 1 week for serum lipid measurements; 3–6 months for atherosclerosis estimation). Administration of test drugs varies, depending on whether prevention or curative effects are being tested. Faeces, if required, are collected and freeze-dried or deep-frozen. They are then homogenised and extracted with chloroform/methanol for analysis. Animals are killed by cervical dislocation and blood collected by cardiac puncture. If rabbits are used, blood monitoring can be carried out at intervals without sacrificing the animal by taking blood samples from the marginal ear vein, as described in the platelet section above. Serum can be stored at − 20 °C until needed.

Triton-induced Hyperlipidaemia   In rats, hypercholesterolaemia can be induced by administering Triton (isooctylpolyoxyethylenephenol), commercially available (e.g. from Sigma), at a dose of 400 mg/kg in saline suspension, administered i.p. They are allowed normal food and water. Measurable hypercholesterolaemia will be established within 18 h. Blood is obtained as above.

Standard Lipid-lowering Agents for Comparison
Clofibrate, 100 mg/kg p.o.
Guggulipid (dose expressed as Z-guggulsterone), 10 mg/kg.

Estimation of Atherosclerosis   Atheromatous lesions are visible and can be measured; the area covered is known to be directly related to circulating lipid levels. Aortas can be removed and fat and connective tissue cleaned off for examination for pathological changes. Atheromatous plaques can be measured planimetrically using a camera lucida, and the area covered expressed as a percentage.

Measurement of Serum Cholesterol   The simplest and most accurate way to measure total cholesterol, LDL-cholesterol, VLDL-cholesterol and HDL-cholesterol is with commercial kits. These need small samples and are now more easily available (e.g. from Sigma). However, standard methods are widely used and can be found in 'Further Reading'. Enzymatic methods are the most commonly used; hydrolysis of cholesterol esters with long-chain fatty acids is necessary, otherwise only free cholesterol would be estimated and this accounts for only about 30% of total cholesterol. The enzymatic methods use cholesterol oxidase to catalyse the reaction of cholesterol with atmospheric oxygen, forming hydrogen peroxide. In the presence of peroxidase the hydrogen peroxide liberates oxygen, which oxidises phenol and antipyrine to give a red colour, which is estimated spectrophotometrically at 500 nm. The method is fairly simple but preparation of reagents less so; they are not very stable. They can, however, be bought in lyophilised form or as stabilised solutions.

Clinical measurements in human volunteers must be compared with normal cholesterol values of 3.9–7.0 mmol/l (150–270 mg/dl). These are of course higher in familial hypercholesterolaemia.

Measurement of Serum Triglycerides Triglycerides are true fats, glycerol esters of long-chain fatty acids. Dietary fats are partially hydrolysed before absorption but reconverted to triglycerides before entering the bloodstream as chylomicrons. They are calculated in serum as triolein; hydrolysis is carried out by transesterification with alkali and heat, or enzymatically, and the glycerol formed estimated colourimetrically.

For all necessary details of these and other lipid determinations, significance and principles involved, *see* 'Further Reading', especially Bauer (1982), *Clinical Laboratory Methods*, 9th Edn.

Related Experiments and Biochemical Determinations

Enzymes Inhibition of the enzyme acyl CoA-cholesterol acyltransferase (ACAT), which is responsible for the conversion of free cholesterol to cholesteryl ester, provides a mechanism through which blood cholesterol levels can be lowered and has been shown to operate when certain dietary indoles are ingested. Plasma lecithin-cholesterol acyltransferase (LCAT) plays a key role in lipoprotein metabolism and an increase in activity is associated with increased levels of HDL. Further detail is outside the scope of this book but can be found in 'Further Reading'.

Hormones Other hormonal parameters which may be measured include testosterone and thyroid hormones, which are found at lower levels with a high fat diet. Testosterone has been found to be decreased in men with myocardial infarction (MI) and thyroid hormones are important regulators of lipid metabolism. Insulin and oestradiol levels have been shown to be increased in patients with MI.

Malondialdehyde (MDA) MDA can be estimated as a metabolite of arachidonic acid formed by platelets during aggregation; platelets are implicated in the early stages of atherogenesis as well as thrombosis.

Liver Homogenates These may be assayed for lipids, cholesterol and mono-oxygenase system parameters during treatment with test substances which may affect metabolic processes in liver microsomes. For example cholesterol 7α-hydroxylase catalyses the conversion of cholesterol to bile acids and is cytochrome P-450-dependent. If the mono-oxygenase system is affected, levels of this may be elevated and changes in NADPH-cytochrome c reductase will occur.

Treatment of Results

Serum Cholesterol, Triglyceride and Lipoprotein–Cholesterol Levels in Diet-induced Hyperlipidaemic Rats

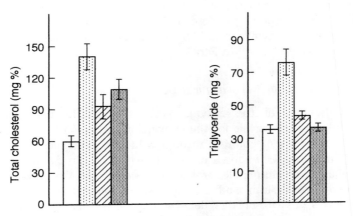

Figure 6.17 Serum cholesterol and triglyceride levels in □, normal, ▩, control (hyperlipidaemic), ▨, bergenin and ▨, guggulip treated rats after 21 days of administration. From Hypolipidaemic activity in rats of bergenin, the major constituent of *Flueggea microcarpa*. Farboodniay Jahromi, M.A. *et al* (1992), *Phytother. Res.*, **6**(4), 180–183, with permission.

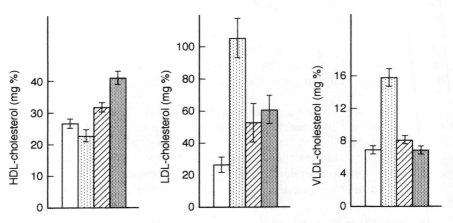

Figure 6.18 Serum HDL-cholesterol, LDL-cholesterol and VLDL-cholesterol levels in □, normal, ▩, control (hyperlipidaemic), ▨, bergenin and ▨, guggulip treated rats after 21 days of administration. From Hypolipidaemic activity in rats of bergenin, the major constituent of *Flueggea microcarpa*. Farboodniay Jahromi, M.A. *et al* (1992), *Phytother. Res.*, **6**(4), 180–183, with permission.

Cholesterol and Total Lipid Level Determination in Triton-induced Hyperlipidaemia in the Rat

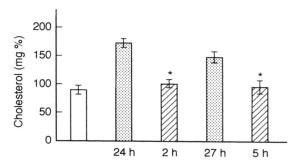

Figure 6.19 Haematic cholesterol levels 24 and 27 h after Triton (100 mg/kg i.p.) and at 2 and 5 h after *Mucuna pruriens* decoction (0.5 g/kg p.o.) administration. Mean ± SE of 10 animals. *$p < 0.05$ compared with Triton. From *Mucuna pruriens* decoction lowers cholesterol and total lipid plasma levels in the rat. Iauk, L. *et al* (1989), *Phytother. Res.*, **3**(6), 263–264, with permission.

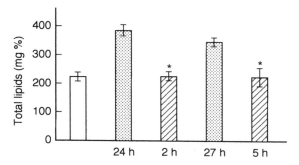

Figure 6.20 Haematic total lipid levels 24 and 27 h after Triton (100 mg/kg i.p.) and at 2 and 5 h after *Mucuna pruriens* decoction (0.5 g/kg p.o.) administration. Mean ± SE of 10 animals. *$p < 0.05$ compared with Triton. From *Mucuna pruriens* decoction lowers cholesterol and total lipid plasma levels in the rat. Iauk, L. *et al* (1989), *Phytother. Res.*, **3**(6), 263–264, with permission.

Lipid Levels in Human Hyperlipidaemic Subjects Given Fenugreek

Table 6.5 Effect of fenugreek seeds on serum lipids

	Cholesterol (mg/dl)		HDL	Ratio of HDL cholesterol		Triglycerides (mg/dl)
	Total	LDL and VLDL		Total cholesterol	LDL and VLDL cholesterol	
I. *Control diet*						
After 20 days	269 ± 6.7	189 ± 8.8	80 ± 3.3	0.29 ± 0.014	0.42 ± 0.031	217 ± 22.3
II. *Experimental diet*						
a) After 10 days	242 ± 7.9[b]	169 ± 9.2[a]	73 ± 7.0	0.31 ± 0.019	0.44 ± 0.042	165 ± 17.1[c]
b) After 20 days	202 ± 5.1[c]	127 ± 6.4[c]	75 ± 4.0	0.37 ± 0.022[b]	0.60 ± 0.065[b]	134 ± 13.6[c]
% Change	−24.4 ± 2.8[c]	−31.7 ± 4.3[c]	5.0 ± 6.9	25.3 ± 8.8[b]	42.3 ± 18.3[b]	−37.7 ± 3.6[c]

Values are mean ± SEM; statistical analysis by paired *t*-test; Control *vs.* Experimental, level of significance
[a] $p < 0.05$; [b] $p < 0.01$; [c] $p < 0.001$. From Hypolipidaemic effect of fenugreek seeds. A clinical study. Sharma, R.D. *et al* (1991), *Phytother. Res.*, 5(3), 145–147, with permission.

Serum Analysis and Measurement of Atheromatous Plaques in Rabbits

Table 6.6 Serum analysis of solasodine-treated rabbits

Group[a]	Total cholesterol (mg/dl)	Triglyceride (mg/dl)	Phospholipid (mg/dl)	HDL cholesterol (mg/dl)
A	106.25 ± 3.60	62.49 ± 4.16	98.5 ± 6.66	43.75 ± 3.60
B	450.00 ± 11.36**	217.50 ± 8.53**	239.0 ± 9.91**	124.00 ± 11.29**
% change	323.58(+)	248.05(+)	142.64(+)	183.42(+)
C	120.25 ± 5.10**	88.04 ± 7.90**	110.42 ± 5.48**	46.80 ± 2.60**
% change	73.26(−)	59.51(−)	70.55(−)	62.25(−)
D	250.00 ± 10.50**, ++	120.83 ± 4.26**, ++	225.50 ± 4.90**, ++	95.5 ± 3.75** ++
% change	44.44(−); (+)107.9	44.45(−); (+)37.24	39.86(−); (+)104.2	23.08(−); (+)104.06

Group[a]	HDL cholesterol Total Cholesterol	VLDL cholesterol (mg/dl)	LDL cholesterol (mg/dl)	Total cholesterol phospholipid	Atherogenic index
A	0.41	12.49 ± 0.83	74.99 ± 3.30	1.07	1.99
B	0.27	43.50 ± 1.70**	343.75 ± 10.95**	1.88	3.12
% change	34.1(−)	248.27(+)	358.39(+)	75.7(+)	
C	0.39	17.79 ± 3.14**	91.24 ± 8.91**	1.09	2.3
% change	44.44(+)	59.10(−)	73.45(−)	42.02(−)	
D	0.38	24.16 ± 1.10***ns	178.66 ± 7.56** ++	1.11	2.12
% change	40.70(+) (−)2.56	44.45(−); (+)35.8	(−)48.62; (+)95.81	40.96(−); (+)1.83	

Values are mean ± SEM, 7 determinations. Groups B compared with group A; group C and D compared with group B: group D compared with group C+. +; $p < 0.01$ Significant *, +; $p < 0.001$ Highly significant **, + +; not significant ns. [a]for key to groups see p. 104.
From Hypocholesterolaemic and anti-atherosclerotic effects of solasodine ($C_{27}H_{42}O_2N$) in cholesterol-fed rabbits. Dixit, V.P. et al (1992). *Phytother. Res.*, 6, 270–273, with permission.

Table 6.7   Faecal analysis of solasodine treated rabbits

Group	Serum cholesterol (mg/dl)	Faecal cholesterol (mg/g)	Serum phospholipid (mg/dl)	Faecal phospholipid (mg/g)
A	106.3 ± 3.6	47.7 ± 1.3	98.5 ± 8.6	32.6 ± 2.0
B	450.0 ± 11.4**	80.7 ± 4.1**	239.0 ± 9.9**	54.5 ± 7.2**
C	120.3 ± 5.1**	97.7 ± 2.2**	110.4 ± 5.5**	67.5 ± 2.2**
D	250.0 ± 10.5**,++	91.5 ± 4.3**, NS	225.5 ± 4.9**,++	61.5 ± 1.2**, NS

Abbreviations and conventions as Table 6.6. From Hypocholesterolaemic and anti-atherosclerotic effects of solasodine ($C_{27}H_{42}O_2N$) in cholesterol-fed rabbits. Dixit, V.P. *et al* (1992), *Phytother. Res.*, 6, 270–273, with permission.

Key to Tables 6.6, 6.7 and 6.8

Group A   Controls.

Group B   Cholesterol fed controls (120 days).

Group C   Cholesterol fed rabbits (60 days) were given solasodine 50 mg/kg OS on days 61–120 through gastric lavage.

Group D   Cholesterol fed rabbits treated with solasodine 50 mg/kg OS on days 1–120.

Table 6.8 Planimetric studies of thoracic and abdominal aorta

Group	Aortic lumen %		Plaque %		Total wall area %	
	Thoracic	Abdominal	Thoracic	Abdominal	Thoracic	Abdominal
A	48.29 ± 2.25	51.02 ± 1.02	NIL	NIL	51.71 ± 4.76	48.97 ± 1.90
B	29.39 ± 0.45**	35.90 ± 1.28**	9.42 ± 0.16	14.60 ± 0.57	70.69 ± 0.45	64.21 ± 1.08
C	41.45 ± 2.79*	44.2 ± 1.40*	3.14 ± 1.98**	3.53 ± 1.20**	67.55 ± 2.75*	55.80 ± 1.80**
D	45.50 ± 1.85**, NS	51.50 ± 1.50**, +	NIL	NIL	54.50 ± 1.75**, +	48.50 ± 1.50**, +

Abbreviations and conventions as Table 6.6. From Hypocholesterolaemic and anti-atherosclerotic effects of solasodine ($C_{27}H_{42}O_2N$) in cholesterol-fed rabbits. Dixit, V.P. et al (1992), *Phytother. Res.*, **6**, 270–273, with permission.

FURTHER READING

Antihypertensives

Methods

Adeagbo, A.S.O. and Okpako, D.T. (1980). Mechanism of noradrenaline potentiation by prostaglandin E_2 in rat mesenteric artery. *Br. J. Pharmac.*, **71**, 75–81.

Arunlakshana, O. and Schild, H.O. (1959). Some quantitative uses of drug antagonists. *Br. J. Pharmacol. Chemother.*, **14**, 48–59.

Calixto, J.B. *et al.* (1992). Vascular action of the crude hydroalcoholic extract from *Hymenaea martiana* on the isolated rat and rabbit aorta. *Phytother. Res.*, **6**(6), 327–331.

Davenport, A.P. and Maguire, J.J. (1994). Is endothelin-induced vasoconstriction mediated only by ET_4 receptors in humans? *Trends Pharm. Sci.*, **15**, 9–11.

Day, M.D. and Dixon, H.V.R. (1971). The action of isoprenaline on the perfused vessels of the rabbits ear. *J. Pharm. Pharmacol.*, **23**, 98–101.

De la Lande, I.S. and Rand, M.J. (1965). A simple isolated nerve-blood vessel preparation. *Aust. J. Exp. Biol. Med. Sci.*, **43**, 639–656.

Ebeigbe, A.B. and Ezimokhai, M. (1988). Vascular smooth muscle responses in pregnancy-induced hypertension. *Trends Pharm. Sci.*, **9**, 455–457.

Friedland, J. and Silverstein, E. (1976). A sensitive fluorimetric assay for serum angiotensin-converting enzyme. *Am. J. Clin. Path.*, **66**(2), 416–424.

Furchgott, R.F. (1983). Role of endothelium in responses of vascular smooth muscle. *Circ. Res.*, **53**, 557–573.

Furchgott, R.F. (1984). The role of endothelium in the responses of vascular smooth muscle to drugs. *Ann. Rev. Pharmacol. Toxicol.*, **24**, 175–197.

Gilani, A.H. and Aftab, K. (1994). Ca^{++} channel blocking activity of *Artemisia scoparia*. *Phytother. Res.*, **8**(3), 145–148.

Hunter, A.J. *et al* (1995). Animal models of acute ischaemic stroke: can they predict clinically successful neuroprotective drugs? *Trends Pharm. Sci.*, **16**, 123–128.

Ichikawa, K. *et al* (1986). The calcium antagonistic activity of lignans. *Chem. Pharm. Bull*, **34**, 3514–3517.

Inokuchi, J.H. *et al* (1984/1985). Inhibitors of angiotensin converting enzyme in crude drugs, I/II. *Chem. Pharm. Bull.*, **32**, 3615–3619 and **33**, 264–269.

Jones, A.W. *et al* (1988). Altered biochemical and functional responses in aorta from hypertensive rats. *Hypertension*, **11**, 627–634.

Kamanyi, A. *et al* (1993). Blood pressure lowering effect of the aqueous extract of the stem of *Ipomoea acanthacarpa* (Convolvulaceae) in spontaneously and salt-loaded hypertensive rats. *Phytother. Res.*, **7**(4), 295–298.

Karaki, H. (1987). Use of tension measurements to delineate the mode of action of vasodilators. *J. Pharm. Methods*, **18**, 1–21.

Kiviranta, J. *et al* (1989). Effects of onion and garlic extracts on spontaneously hypertensive rats. *Phytother. Res.*, **3**(4), 132–135.

Laragh, J.H. and Brenner, B.M. Eds. (1990). The discovery and physiological effects of a new class of highly specific angiotensin-II receptor antagonists. In *Hypertension: Pathophysiology, Diagnosis, and Management*, pp. 2351–2360, Raven Press, NY.

McGregor, D.D. (1965). The effect of sympathetic nerve stimulation on vasoconstrictor responses in perfused mesenteric blood vessels of the rat. *J. Physiol.*, **177**, 21–30.

Meunier, M.T. *et al* (1987). Inhibitors of angiotensin-I converting enzyme by flavonolic compounds: *in vitro* and *in vivo* studies. *Planta Medica*, **53**, 12–15.

Moncada, S. *et al* (1986). Mechanism of action of endothelium-derived relaxing factor. *Proc. Natl Acad. Sci. USA*, **83**, 9164–9168.

Moncada, S. *et al* (1991). Nitric oxide: physiology, pathophysiology and pharmacology. *Pharmacol. Rev.*, **43**, 109–142.

Olesen, J. *et al* (1994). Nitric oxide is a key molecule in migraine and other vascular headaches. *Trends Pharm. Sci.*, **15**, 149–153.

Rauwald, H.W. *et al* (1994). Screening of nine vasoactive plants for their possible calcium antagonistic activity. *Phytother. Res.*, **8**(3), 135–140.

Sanchez de Rojas, V.R. *et al* (1994). Pharmacological activity of the extract of *Satureja obovata* varieties on isolated smooth muscle preparations [includes influence of endothelium]. *Phytother. Res.*, **8**(4), 212–217.

Schultz, R. and Triggle, C. (1994). Role of NO in vascular smooth muscle and cardiac muscle formation. *Trends Pharm. Sci.*, **15**, 255–259.

Victor, J.D. (1977). Significance of the vascular renin–angiotensin pathway. *Hypertension*, **8**, 553–559.

Yamahara, J. *et al* (1989). The effect of alismol isolated from *Alismatis rhizoma* on experimental hypertensive rats. *Phytother. Res.*, **3**(2), 57–60.

Reviews

Hunter, A.J. *et al* (1995). Animal models of acute ischaemic stroke: can they predict clinically successful neuroprotective drugs? *Trends Pharm. Sci.*, **16**, 123–128.

Karaki, H. and Weiss, G. (1988). Calcium release in smooth muscles. *Life Sci.*, **42**, 111–122.

Linz, W. *et al*. (1995). Contribution of kinins to the cardiovascular actions of angiotensin-converting enzyme inhibitors. *Pharm. Rev.*, **47**, 25–49.

Spedding, M. and Cavero, I. (1984). Calcium antagonists: a class of drugs with a bright future. Part II: determination of basic pharmacological properties. *Life Sci.*, **35**, 575–587.

Van Breeman, C. *et al* (1982). Selectivity of calcium antagonist action in vascular smooth muscle. *Am. J. Cardiol.*, **49**, 547–553.

Direct Action on the Myocardium

Methods

Burns, J.H. (1952). *Practical Pharmacology* (p. 52). Blackwell, Oxford.

Clark, C. *et al* (1980). Coronary artery ligation in anaesthetised rats as a method for the production of experimental dysrhythmias, and for the determination of infarct size. *J. Pharmacol Methods*, **3**, 357–368.

Friedrichs, G.S. *et al* (1994). Antifibrillatory effects of *Clofilium* in the rabbit isolated heart. *Br. J. Pharmacol.*, **113**, 209–215.

Kimura, M. *et al* (1989). Positive inotropic action and conformation difference of lupin alkaloids in isolated cardiac muscle of guinea-pig and bullfrog. *Phytother. Res.*, **3**(3), 101–105.

Kimura, I. *et al* (1994). Aconitine-induced bradycardia, centrally acting muscarinic effects are inhibited peripherally by higenamine in conscious mice. *Phytother. Res.*, **8**(3), 129–134.

Limaye, D.A. *et al* (1995). Cardiovascular effects of the aqueous extract of *Moringa pterygosperma*. *Phytother. Res.*, **9**, 37–40.

Marshall, R.J. and Parratt, J.R. (1975). Anti-arrhythmic, haemodynamic and metabolic effects of 3α-amino-5α-androstan-2β-ol-17-one hydrochloride in greyhounds following acute coronary ligation. *Br. J. Pharmacol.*, **55**, 359–368.

McLeod, D.P. and Reynolds, A.K. (1962). Studies on some rauwolfia preparations with special reference to their effects on cardiac arrythmias. *Arch. Intern. Pharmacodynamics*, **138**, 347–350.

Narula, O.S. *et al* (1978). A new method for measurement of sinoatrial conduction time. *Circulation*, **58**, 706.

Occhiuto, F. *et al*. (1991). Comparative anti-arrhythmic and anti-ischaemic activity of some flavones in the guinea-pig and rat. *Phytotherapy Res.*, **5**(1), 9–14.

Radhakrishnan, R. *et al* (1993). *Terminalia arjuna*, an Ayurvedic cardiotonic, increases contractile force of rat isolated atria. *Phytother. Res.*, **7**(3), 266–268.

Scherf, D. (1947). Studies on auricular tachycardia caused by aconitine administration. *Proc. Soc. Exp. Biol. Med.*, **64**, 233–239.

Sinha, J.N. *et al* (1971). Centracenic cardiac arrhythmia induced by aconitine: a new 'model' for screening of anti-arrhythmic drugs. *Jpn. J. Pharmacol.*, **21**, 699–706.

Turner, R.A. (1965). Antifibrillatory agents (Chapter 19); Cardiotonics (Chapter 20). In *Screening Methods in Pharmacology*, Academic Press, London.

Key Papers and Reviews

Brodde, O.-E. (1991). β_1- and β_2-adrenoceptors in the human heart: properties, function and alterations in chronic heart failure. *Pharmacol. Rev.*, **43**(2), 203–242.

Erhardt, P.W. *et al* (1987). Cardiotonic agents. 3. A topographical model of the cardiac cAMP phosphodiesterase receptor. *Molec. Pharmacol.*, **33**, 1–13.

Harrison, D.C. (1986). Current classification of anti-arrhythmic drugs as a guide to their rational clinical use. *Drugs*, **31**, 93–95.

Kaufmann, A.J. (1994). Do human atrial 5-HT$_4$ receptors mediate arrhythmias? *Trends Pharm. Sci.*, **15**, 451–455.

Nattel, S. (1991). Anti-arrhythmic drug classifications: a critical appraisal of their history, present status, and clinical relevance. *Drugs*, **41**, 672–701.

Terzic, A. *et al* (1993). Cardiac α_1-adrenoceptors: an overview. *Pharmacol. Rev.*, **45**(2), 147–176.

Thomas, R. *et al* (1990). Digitalis: its mode of action, receptor and structure–activity relationships. *Adv. Drug Res.*, **19**, 311–362.

Touboil, P. and Waldo, A.L., Eds. (1990). *Atrial arrhythmias*. Mosby, St Louis.

Vaughan Williams, E.M. (1975). Classification of antidysrhythmic drugs. *Pharmacol. Ther.*, **1**, 115–138.

Antiplatelet Activity

Methods Using Platelet Aggregation

Azuma, H. *et al* (1986). Endothelium-dependent inhibition of platelet aggregation. *Br. J. Pharmacol.*, **88**, 411–415.

Beretz, A. and Cazenave, J.-P. (1991). Platelet aggregation methods. In *Methods in Biochemistry, Vol. 6. Assays for Bioactivity*, Ed. K. Hostettman. Academic Press, London.

Born, G.V.R. (1962). Aggregation of blood platelets by adenosine diphosphate and its reversal. *Nature*, **194**, 927–929.

Colman, R.W. and Smith, J.B., Eds. (1987). *Methods for Studying Platelets and Megakaryocytes*, CRC Press, Boca Raton, CA.

Harker, L.A. and Zimmermann, T.S., Eds. (1983). *Measurements of Platelet Function. Methods in hematology*, Vol. 8, Churchill-Livingstone, Edinburgh.

Kinlough-Rathbone *et al* (1977). Mechanism of platelet shape change, aggregation with release induced by collagen, thrombin or A23187. *J. Lab. Clin. Med.*, **90**, 707–719.

Radomski, M. and Moncada, S. (1983). An improved method for washing of human platelets with prostacyclin. *Thromb. Res.*, **30**, 383–389.

Zucker, M.B. (1989). Platelet aggregation measured by the photometric method. *Methods Enzymol.*, **169**, 118–133.

Thrombosis Tests in vivo

Beckemeier, H. *et al* (1984). Carageenan-induced thrombosis in the rat and mouse as a test model of substances influencing thrombosis. *Biomed. Biochim. Acta*, **43**, S347–S350.

Guarneri, L. *et al* (1988). A new model of pulmonary microembolism in the mouse. *J. Pharmacol. Meth.*, **20**, 161–167.

Philp, R.B. (1981). *Methods of Testing Proposed Anti-thrombotic Drugs*, CRC Press, Boca Raton, CA.

Reviews

Colman, R.W. *et al* Eds. (1987). *Haemostasis and Thrombosis, Basic Principles and Clinical Practice*, J.B. Lippincott, Philadelphia, PA.

Holmsen, H., Ed. (1986). *Platelet Responses and Metabolism, Vol. I. Responses.* CRC Press, Boca Raton, CA.

Hourani, S.M.O. and Hall, D.A. (1994). Receptors for ADP on human blood platelets. *Trends. Pharm. Sci.*, **15**, 103–108.

Weksler, B.B. (1983). Platelets and the inflammatory response. *Clin. Lab. Med.*, **3**, 667–677.

Mechanistic Studies

Guarneri, L. *et al* (1990). Genistein, a protein tyrosine kinase inhibitor, inhibits thromboxane A_2-mediated human platelet response. *Mol. Pharmacol.*, **39**, 475–480.

Williamson, E.M. *et al* (1981). Studies on the mechanism of action of 12-deoxyphorbolphenylacetate. *Biochem. Pharmacol.*, **30**(19), 2691–2696.

Cyclo-oxygenase Assays

Samuelsson, B. *et al* (1981). *Advances in Prostaglandin and Thromboxane Research*, Vol. 6, Raven Press, New York.

Antisclerotic Testing, Lipid and Cholesterol-lowering

(*See also* references quoted in 'Treatment of Results' section, and cited in review papers.)

General: Hyperlipidaemia and Coronary Heart Disease

Anderson, K.M. *et al* (1987). Cholesterol and mortality: 30 years of follow-up from the Framingham study. *JAMA*, **257**, 2176–2180.
Beaumont, J.L. *et al* (1970). Classification of hyperlipidaemias and hyperlipoproteinaemias. *Bull. WHO*, **43**, 891–915.
Miller, G.J. and Miller, N.E. (1975). Plasma high density lipoprotein concentration and the development of ischaemic heart disease. *Lancet*, **1**, 16–19.
Patsch, W. and Gotto, A. Jr (1995). High density lipoprotein cholesterol, plasma triglyceride, and coronary heart disease: pathophysiology and management. *Adv. Pharmacol.*, **32**, 375–426.
Segal, P. *et al* (1984). Lipids and dyslipoproteinemia. In *Clinical Diagnosis and Management.* Ed. J.B. Henry, pp. 180–302. Saunders, Philadelphia, PA.
Sliskovic, D.R. and White, A.D. (1991) Therapeutic potential of ACAT inhibitors as lipid lowering and antiatherosclerotic agents. *Trends Pharm. Sci.*, **12**, 194–199.

Clinical Studies

Madar, Z. *et al* (1988). Glucose-lowering effect of fenugreek in non-insulin dependent diabetics. *Eur. J. Clin. Nutr.*, **42**, 51–54.
Mendoza, S.G. *et al* (1983). Estradiol, testosterone, apolipoproteins, lipoprotein cholesterol and lipolytic enzymes in men with premature myocardial infarction and angiographically assessed coronary occlusion. *Artery*, **12**, 1–23.
Swain, J.F. *et al* (1990). Comparison of the effects of oat bran and low-fibre wheat on serum lipoprotein levels and blood pressure. *N. Engl. J. Med.*, **322**, 147–152.

Methods

Estimation of Cholesterol, Triglycerides, etc.

Allain, C.C. *et al* (1974). Enzymatic determination of total serum cholesterol. *Clin. Chem.*, **20**, 470–475.
Bauer, J.D. (1982). *Clinical Laboratory Methods*, 9th Edn. C.V. Mosby Co., St Louis, MI.
Bergmeyer, H.U., Ed. (1974). *Methods of Enzymic Analysis*, Vol. 4, pp. 1831–1835. Verlag Chemie Weinheim Academic Press, London and New York.
Burstein, M. *et al* (1970). Rapid method of isolation of lipoprotein from human serum by precipitation of polyanion. *J. Lipid Res.*, **11**, 583–587.
Freidward, W.T. *et al* (1972). Estimation of concentration of low density lipoprotein cholesterol in plasma without the use of preparative ultracentrifuge. *Clin. Chem.*, **19**, 449–452.
Frings, C.S. and Dunn, R.T. (1970). A colorimetric method for determination of the total serum lipids based on the sulpho-phospho-vanillin reaction. *Am. J. Clin. Pathol.*, **53**, 89–91.
Gottfreid, S.P. and Rosenberg, B. (1973). Improved manual spectrophotometric procedure for determination of serum triglyceride. *Clin. Chem.*, **19**, 1077–1078.
Patch, W. and Gotto, A.M. Jr (1995). High density lipoprotein, cholesterol, plasma triglyceride and coronary heart disease. In: *Adv. Pharmacol.*, **32**, 375–426. Academic Press, NY.
Rudel, L.L. and Morris, M.D. (1973). Determination of cholesterol using *o*-phthalaldehyde. *J. Lipid Res.*, **14**, 364–368.
Zilversmit, D.B. and Davis, A.K. (1950). Microdetermination of plasma phospholipid by trichloroacetic acid precipitation method. *J. Lab. Clin. Invest.*, **35**, 155–160.

Zlatkis, A. *et al* (1953). A method for the determination of serum cholesterol. *J. Clin. Med.*, **41**, 486–492.

Diet-induced Hyperlipidaemia

Bishayee, A. and Chatterjee, M. (1994). Hypolipidaemic and anti-atherosclerotic effects of oral *Gymnema sylvestre* leaf extract in albino rats fed on a high fat diet. *Phytother. Res.*, **8**(2), 118–120.

Dunn, S.E. and LeBlanc, G.A. (1994). Hypocholesterolemic properties of plant indoles. *Biochem. Pharmacol.*, **47**(2), 359–364.

Forestieri, A.M. *et al* (1989). Effects of guar and carob gums on glucose, insulin and cholesterol plasma levels in the rat. *Phytother. Res.*, **3**(1), 1–4.

Kimura, Y. *et al* (1987). Effects of Japanese and Chinese traditional medicines 'Hachimi-Gan' (Ba-Wei-Wan) on lipid metabolism in rats fed high sugar diet. *Planta Medica*, **53**(2), 128–131.

Triton-induced Hyperlipidaemia

Schurr, P.E. *et al* (1972). Triton-induced hyperlipidemia in rats as an animal model for screening hypolipidemic drugs. *Lipids*, **7**, 68–74.

Significance and Measurement of Related Parameters

Wojcicki, J. *et al* (1988). Effect of Padma 28 on experimental hyperlipidaemia and atherosclerosis induced by high fat diet in rabbits. *Phytother. Res.*, **2**(3), 145–147.

Determination of LCAT

Nagasaki, T. and Akanuma, Y. (1977). A new colorimetric method for the determination of plasma lecithin–cholesterol acyltransferase. *Clin. Chem. Acta*, **75**, 371–375.

Malondialdehyde, Involvement and Determination

MacFarlane, D.E. *et al* (1977). Malondialdehyde production by platelets during secondary aggregation. *Thromb. Haemost.*, **38**, 1002–1009.

Stuart, M.J. *et al* (1975). A simple method non-radio-isotope technique for the determination of platelet life-span. *N. Engl. J. Med.*, **292**, 1310–1313.

7

The Respiratory System

The most important disorders of the respiratory system, excluding infections, are the allergic conditions asthma and hayfever. For various reasons, mostly unknown, the prevalence of these is increasing. Both are caused by the same basic processes: IgE antibodies attach to mast cells and renewed exposure to antigen causes degranulation of mast cells; also, inflammatory damage to the endothelium of the airways causes the underlying muscle to become super-sensitive to external irritants and inflammatory mediators. These include *histamine, leukotrienes, thomboxanes, platelet activating factor (PAF)* and chemo-tactic agents. If this occurs mainly in the nose, eyes and throat, the result is hayfever. When the airways are involved, with bronchospasm, mucosal oedema and excessive mucus production, leading to coughing and apnoea (wheezing), the result is asthma. Rarely, massive release of mediators leads to anaphylaxis, which is a life-threatening condition triggered by allergens such as bee stings or drugs. It involves bronchospasm, oedema in the nose and throat, and cardiovascular collapse and the treatment is with intrave-nous adrenaline, antihistamines and corticosteroids.

Asthma is much more common but still serious, and may be classified into extrinsic asthma, caused by specific allergens, and intrinsic asthma, where there is no obvious allergic cause. Acute severe attacks are known as status asthmaticus. The clinical management of acute asthma at present is with oxygen if necessary; then bronchodilators: β_2 receptor agonists, e.g. salbuta-mol, xanthines, e.g. theophylline and aminophylline, and antimuscarinics, e.g. the atropine derivative ipratropium bromide; and anti-inflammatory therapy with corticosteroids. Mast cell stabilisers are very important in pro-phylactic treatment, the main one being sodium cromoglycate. The story of this compound is a fine example of serendipity, occurring during research on khellin, a naturally occurring chromone compound extracted from the tooth-pick plant, *Ammi visnaga*. Khellin derivatives relax guinea-pig bronchial smooth muscle, but cromoglycate also protects the animals from the effects of bronchoconstrictors such as histamine and acetylcholine. Its mechanism of action is unclear, but it reduces the influx of calcium into antigen-sensitised mast cells and prevents the release of histamine.

Novel drug development is targeted also on the inflammatory mediators already mentioned; however, the pathways involved are of fairly recent discovery and research on these lines in its infancy. Natural products are also well represented in current asthma research, e.g. the PAF antagonists ginkgolide B (BN 52021), from *Ginkgo biloba*, and the lignan kadsurenone from *Piper futokadsurae*.

The experimental section following will be devoted to testing for anti-asthmatic activity, since asthma is so important and there is a tremendous amount of new information available. Hayfever can be treated with similar drugs, taken systemically or applied locally to the nose and eyes.

Antitussives are widely used despite difficulties with proving efficacy; they will be discussed briefly. No new methods of testing have been developed recently so these will be referenced only.

TESTING FOR ANTI-ASTHMATIC ACTIVITY

The inflammatory mediators most under scrutiny for development of therapeutic antagonists, because of their involvement with asthma, are PAF and the leukotrienes. These will be discussed in more detail here. However, the methods given, i.e. isolated bronchial smooth muscle or sensitised animals experiments, are suitable for testing extracts as bronchodilators or inflammatory antagonists, regardless of their mechanism of action (anticholinergic, antihistamine, β_2 agonist, PAF antagonist, etc.). Cyclo-oxygenase products of arachidonic acid metabolism appear to be less important in the aetiology of asthma than the lipoxygenase products, the leukotrienes. In fact, cyclo-oxygenase inhibitors are known to be able to precipitate an attack of asthma, possibly by diverting substrate (arachidonic acid) to lipoxygenase pathways and producing bronchoconstricting leukotrienes and lipotoxins. Indomethacin, which suppresses cyclo-oxygenase in the lung, reduces concentrations of PGE_2, a bronchodilator which also inhibits further release of inflammatory mediators. The reduction in PGE_2 levels therefore enhances the response to bronchoconstrictors. *See* Figure 7.1 and 'Methods' section for clarification.

PAF Antagonists

PAF (platelet activating factor) is a pro-inflammatory mediator with effects in many body systems as well as the lung. It is particularly important in the cellular immune response; with involvement in auto-immune diabetes mellitus, in allograft rejection, in renal immune injury, atopic eczema, cerebral ischaemia and immune keratitis. It is released during treatment with the immunosuppressant cyclosporin A. This widespread activity must be borne in mind when investigating PAF antagonists, which are at present being targeted for asthma, and there is potential for PAF antagonists in other immune disorders.

In the lung, PAF has been shown to activate the arachidonic acid cascade with a dose-dependent generation of thromboxane. Formation of PAF in lungs may also contribute to histamine release in anaphylaxis. It is now apparent that many drugs of long-standing clinical application in asthma have modulating effects on PAF activity or generation; e.g. aminophylline, a xanthine bronchodilator, has been shown to antagonise PAF-induced bronchoconstriction; and sodium cromoglycate affects the cellular response to PAF *in vitro* and *in vivo*, as does atropine.

The most important and clinically advanced PAF antagonists at present are of natural origin. They are divided into three groups:

1. Ginkgolides.
2. Lignans.
3. Gliotoxins.

Ginkgolides

These are C_{20} diterpenes, described in the literature as BN52063 (a mixture of ginkgolides) and BN52021 (ginkgolide B). These have been extensively investigated and reviewed by Braquet *et al.* (1987) and others (*see* 'Further Reading') and shown to suppress PAF-induced bronchoconstriction and airway hyper-reactivity in humans and animals. They exert a protective effect against antigen-induced bronchial provocation tests in asthmatic patients.

Lignans

The neolignan kadsurenone was the first natural product discovered as a potent PAF inhibitor, using rabbit platelets. It also inhibits degranulation of human neutrophils and antagonises other PAF-induced phenomena. Other lignans of interest as PAF antagonists include burseran, from *Bursera microphylla*, and the nectandrins A and B, from *Nectandra rigida*. *See* review articles cited in 'Further Reading' for further information.

Gliotoxins

These are epipolythiodioxo piperazines (ETPs) produced by a wide range of fungi. Gliotoxin itself is a relatively weak PAF inhibitor but various derivatives have been shown to inhibit PAF-induced bronchoconstriction in rats. *See* 'Further Reading'.

5-Lipoxygenase Inhibitors

Lipoxygenase enzymes catalyse the oxidative metabolism of a range of unsaturated fatty acids. In mammalian systems the substrate is usually arachido-

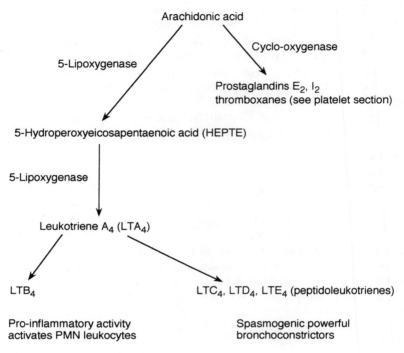

Figure 7.1   5-Lipoxygenase pathway

nic acid, and 5-lipoxygenase produces the potent biological mediators the leukotrienes (LTs). The pathway is shown in Figure 7.1.

Leukotriene B_4 (LTB$_4$) is a potent chemotactic agent; it induces polymorphonuclear leukocyte migration *in vivo* and induces inflammatory reactions in man. It is also found in elevated concentrations in the synovial fluid of patients with rheumatoid arthritis, in rectal dialysis fluid from colitic patients, in psoriatic skin and in other auto-immune inflammatory disorders. In asthma it appears that the lipoxygenase metabolites, the peptidoleukotrienes LTC$_4$ LTD$_4$ and LTE$_4$, are the most important as they have been shown to be powerful spasmogens in airway smooth muscle and vasculature. The slow-reacting substance of anaphylaxis, formerly known as SRS-A, is now known to be a mixture of LTD$_4$ and LTE$_4$. In fact these mediators are at least 100 times more potent than histamine or methacholine (an analogue of acetylcholine) as bronchoconstrictors. Asthmatic patients show enhanced sensitivity to these agents and therefore inhibition of 5-lipoxygenase is considered to be a possible route to the development of new anti-asthma agents. There are some naturally occurring lipoxygenase inhibitors, such as the flavonoids baicalein and quercetin, but none at a clinical stage yet.

β_2-Adrenoceptor Agonists

β-Adrenergic receptors (adrenoceptors) are divided into two types, β_1 and β_2. β_1 receptors are involved in sympathomimetic activity on the heart, whereas β_2 receptors are concentrated in the lung and are involved in bronchodilation. β_2 agonists such as salbutamol are widely used in the treatment of asthma, especially during acute attacks where the airways must be opened immediately, despite fears that overuse is leading to airway hyper-reactivity and toxicity problems.

If β_2 activity is to be investigated, this can be done using bronchial or tracheal smooth muscle, and antagonised using a β-blocker. β-agonist activity may also be found during experiments on various types of smooth muscle (*see* Chapter 6; 'The Cardiovascular System').

β-blockers are drugs used to treat angina and hypertension, by blocking β_1 activity. If they antagonise only β_1 receptors, they are termed cardioselective; however, if they also antagonise β_2 activity they may precipitate an asthma attack. For this reason most β-blockers are avoided for use in asthmatics. Compounds which block β_2 as well as β_1 receptors include propranolol, timolol and nadolol. Some β-blockers, such as propranolol, also affect enzymes such as phospholipase A_2, which must be taken into consideration when interpreting results.

Animal Models: anti-asthmatics

The following *in vitro* experiments will be described: contraction of tracheal spirals and lung parenchyma strips with various agonists; and *in vivo* bronchoconstriction in an anaesthetised animal.

Rats and guinea-pigs are used. The rat produces a reaginic antibody, IgE, which is rather similar to that produced by human asthmatics. Rat lung is more similar to human lung in pharmacological response, but guinea-pig lung resembles it more closely histologically. It is usually a good idea to use both animals. Sensitised preparations are particularly useful; the animal is pretreated with the antigen ovalbumin, and when tissues are later removed, contractions and bronchoconstrictions can be induced by stimulating with the same antigen.

IN VITRO METHODS

Tracheal Spirals and Lung Parenchyma Strips

The tracheal spiral (Table 7.1) is often used to represent the large airways; it is a useful model although fairly difficult to prepare. Lung parenchyma represents peripheral airways. These are usually taken from the same animal. Contractions can be induced with histamine, carbachol (an analogue of acet-

Table 7.1   Use of tracheal spirals from sensitised animals

Tissue	Agonist	Mediators
Non-sensitised	Arachidonic acid	Mainly cyclo-oxygenase products
Non-sensitised + indomethacin	Arachidonic acid	Lipoxygenase products (LTs)
Sensitised	Antigen (ovalbumin)	Arachidonic acid metabolites: cyclo-oxygenase and lipoxygenase products
Sensitised + indomethacin	Antigen	ENHANCED RESPONSE to LTs, PAF, histamine

This scheme applies also to lung parenchyma strips, with the difference that the lipoxygenase pathway is less important, and the enhanced response in indomethacin pre-treated sensitised tissue is not seen, and cannot be used specifically to test for leukotriene mediators.

ylcholine), LTD_4, PAF and arachidonic acid, and the effect of plant extracts tested against these directly.

Arachidonic acid-induced contractions are used to evaluate both cyclo-oxygenase and lipoxygenase modulators. In the presence of indomethacin, a cyclo-oxygenase inhibitor, the lipoxygenase pathway predominates and drugs which affect leukotriene-induced contractions can be evaluated.

When tissues from sensitised animals are used, they respond to other mediators in addition to arachidonic acid metabolites. If sensitised tissues are pre-treated with indomethacin, the cyclo-oxygenase pathway is suppressed and, with it, production of one of its products, PGE_2. The latter is itself a bronchodilator, and it also inhibits the further release of other inflammatory mediators, including arachidonic acid. It has been shown that removing the influence of PGE_2 leads to amplified release of arachidonic acid and an increased contractile response of the tissue. Therefore, in indomethacin-pre-treated tissues from sensitised animals, antigen challenge leads to contractile responses, mediated through peptidoleukotrienes and other non-cyclo-oxygenase products such as PAF and histamine.

IN VIVO METHODS

In Vivo *Bronchoconstriction: Dixon and Brodie Technique*

An anaesthetised animal can be used to measure bronchoconstriction when the trachea is connected to a pressure transducer and the animal artificially ventilated. Drugs can be given orally or by injection and their effect on airflow monitored. Again, sensitised animals may be used and bronchoconstriction induced by antigen stimulation. In this case bronchoconstriction starts about 1 min after antigen challenge, depending on dose, and rapidly reaches a maximum.

Experiments Used in Conjunction with the Above

1. If PAF antagonism is suspected, platelet aggregation methods are suitable. Use rabbit platelets and the method described in the platelet section. PAF-acether (approx. 2.5 nM) can be used as the agonist, and the platelets pretreated with plant extract. Ginkgolide B (BN52021) can be used for comparison.
2. If antihistamine activity may be involved, the guinea-pig ileum preparation outlined in Chapter 4, 'The Gastro-intestinal System' can be employed, with histamine as the agonist.

Materials and Methods: anti-asthmatics (Figures 7.2–7.5; Tables 7.2–7.3)

Animals

Guinea-pigs of 400–500 g body weight for all experiments; rat for lung parenchyma and *in vivo* experiments.

Reagents

	Dose ranges	
Agonist	*in vitro*	*in vivo*
Agonists/bronchoconstrictors causing contraction of airway smooth muscle		
Arachidonic acid	20 µg–50 µg/ml	
Leukotrienes		
LTB$_4$	10^{-12}–10^{-9}M	
LTD$_4$	10^{-9}–10^{-7}M	
Histamine	10^{-6}–10^{-4}M	
Carbachol	10^{-8}–10^{-4}M	
PAF	3 ng	60–100 ng/kg
Ovalbumin (grade V)	10 µg/ml	0.75 mg/kg
Inhibitors and bronchodilators		
Isoprenaline:		
β-receptor agonist, bronchodilator	10^{-9}–10^{-6}M	
Ginkgolide B (BN 52021):		
PAF antagonist		20 mg/kg p.o.
		60–100 ng/kg i.v.
Indomethacin:		
cyclo-oxygenase inhibitor	10^{-6}–10^{-7} in organ bath	
Plant extracts to test		500 mg/kg p.o. to start
Isolated compounds to test		50 mg/kg p.o. to start

Sensitisation of Guinea-pigs

Animals are actively sensitised by injecting with ovalbumin (100 mg s.c. or i.p.), 3–4 weeks before experiments. Antisera, such as rabbit antichicken ovalbumin antiserum (1 ml/kg of a 1:4 dilution), can be used for passive sensitisation and experiments performed 24 h later.

Methods

Tracheal Spiral   Lungs and trachea are removed immediately after the animal has been killed by cervical dislocation, and bled from the aorta. The organs are transferred to cold Krebs solution (*see* Appendix II). The trachea is carefully cut in a spiral (like a helical spring) so that when extended under tension its length will contract and relax under the influence of the agonists and antagonists administered. Four segments can be obtained from a single trachea. These are each suspended in a 10 ml organ bath containing Krebs solution maintained at 37 °C and aerated with 95% O_2 and 5% CO_2 under a tension of 1 g. This is connected to a pressure transducer and recorder in the usual way. The preparations are allowed to equilibrate for at least 1 h until a constant resting tone is obtained.

The four tissue segments can be used for comparative purposes, e.g. one as a control (vehicle only) and three for various concentrations of test plant extracts against one agonist. Responses to histamine may be used to test the efficacy of the preparation. If pretreatment with indomethacin is required, this is carried out at least 30 min before the addition of agonists and test materials.

Responses to test extracts can be measured as simple contractions, and expressed as a percentage of the maximal contraction obtained, either to the same agonist or relative to that of histamine (*see* examples in 'Treatment of Results'). In some instances a form of normalisation is carried out by dividing responses throughout the experiment by the tissue wet weight at the end. Cumulative concentration–response curves may then be drawn.

Relaxation of tissue may be measured by first contracting with histamine or other agonist, then relaxing with plant extract or a standard bronchodilator such as isoprenaline. A relaxation curve (e.g. percentage relaxation relative to isoprenaline maximum) against log dose may then be plotted.

Lung Parenchyma Strips   Tissues are removed from the animal as above. Strips of parenchyma are cut from the distal edges of the lobes of the lung and each suspended in a 10 ml organ bath as described, with a 500 mg tension. Four preparations can be taken from each animal and used as before. If the agonist used for contraction is PAF, strips are used once only, since PAF induces tachyphylaxis.

In vivo *Bronchoconstriction in Guinea-pig* The guinea-pig is anaesthetised (ethyl carbamate 1.5 mg/kg or other suitable method), tracheotomised and artificially ventilated with a respiratory pump (2-way, 70–80 strokes/min, 1 ml/100 g/stroke). A pneumothorax is performed to eliminate spontaneous breathing. The initial resistance is kept constant (e.g. 10 cm water) and excess air measured with a bronchospasm transducer connected to a recorder. Drugs may be administered orally before the experiment or via a catheter into a vein (usually the jugular). Bronchoconstrictors such as PAF are introduced i.v. in this way.

Bronchoconstriction is expressed as a percentage of the maximum obtained when the trachea is clamped at the end of the experiment. In sensitised animals it is expressed as a percentage of the increase in resistance over the base line. Inhibition in both cases is calculated as the percentage difference in average bronchoconstriction between control and treated animals.

Antitussive testing

The cough reflex is a necessary physiological mechanism to clear the respiratory passages where there is excessive secretion in conditions such as bronchitis. Therefore, it is not always appropriate to suppress a cough. Many coughs are now thought to be due to asthma where there is no infection present, and it would be more reasonable to treat this with a bronchodilator than with an antitussive. However, the general public firmly believe in cough medicines and they are useful for treating irritating coughs associated with colds and bronchitis.

There are several animal experiments described in the literature and these are referenced in 'Further Reading'. As none is of recent development they will not be detailed here. With clinical studies in human volunteers it has been found that correlations between subjective assessment by the patient, and actual evidence of cough inhibition using a tape recorder, are poor. Most tests use citric acid as a harmless cough inducer in humans and guinea-pigs; but ammonia has been used as a tussogen in the cat, and dilute sulphuric acid in the dog. Codeine is usually used as a standard antitussive. So far most cough suppressants are of opioid derivation.

Treatment of Results

Tracheal Spirals from Non-sensitised Guinea-pigs

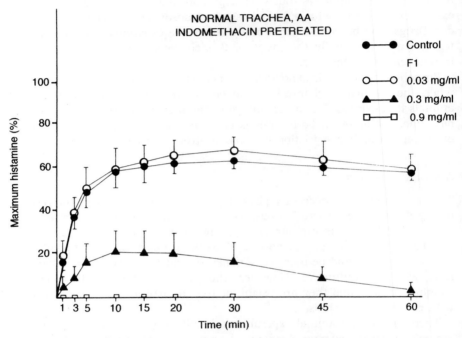

Figure 7.2   Effect of different concentrations of F1 on arachidonic acid-induced contraction of indomethacin-pretreated tracheal spirals obtained from non-sensitised guinea-pigs. AA = arachidonic acid; F1 = fraction 1 of plant extract. From Dose–response effects of one subfraction of *Desmodium adscendens* aqueous extract on antigen- and arachidonic acid-induced contractions of guinea-pig airways. M.E. Addy and J.F. Burka (1987), *Phytother. Res.*, 1(4), 180–186, with permission.

Sensitised Guinea-pigs

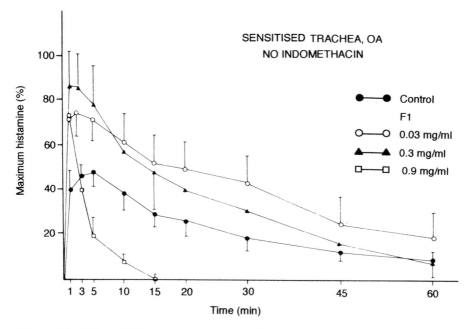

Figure 7.3 Effect of different concentrations of F1 on ovalbumin-induced contractions of tracheal spirals from actively sensitised guinea-pigs. OA = ovalbumin; F1 = fraction 1 of plant extract. From Dose–response effects of one subfraction of *Desmodium adscendens* aqueous extract on antigen- and arachidonic acid-induced contractions of guinea-pig airways. M.E. Addy and J.F. Burka (1987), *Phytother. Res.*, **1**(4), 180–186, with permission.

Tracheal Spirals: Relaxation of Histamine-induced Contractions

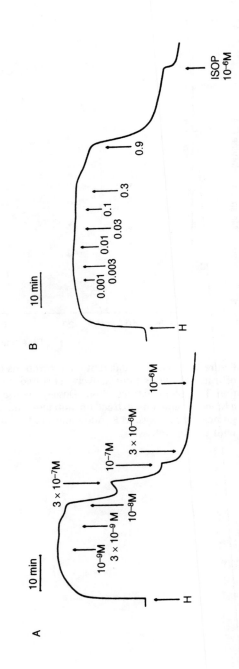

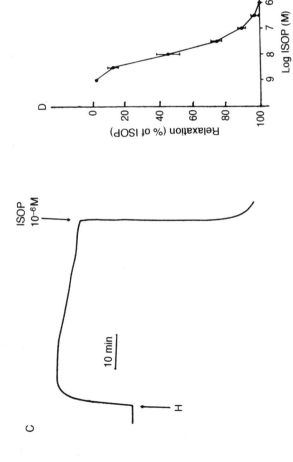

Figure 7.4 Relaxation of histamine-contracted tracheal spirals in response to cumulative concentrations of isoprenaline (A); DAF1 (B); and vehicle (C). Cumulative concentration–response curves for isoprenaline and DAF1 are shown in (D). $n = 4$ for (D). Values in (B) are mg/ml. From Effect of *Desmodium adscendens* fraction F1 (DAF1) on tone and agonist-induced contractions of guinea-pig airway smooth muscle. M.E. Addy and J.F. Burka (1989), *Phytother. Res.*, 3(3), 85–90, with permission.

Guinea-pig Lung Parenchyma Strips (GPLP): Effect of Gossypol on Contractions Induced by Various Agonists

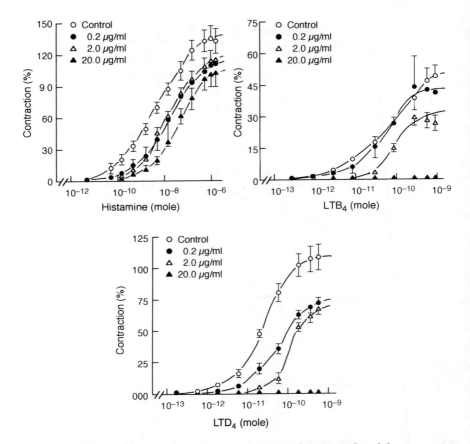

Figure 7.5 Effects of gossypol on the contractions of GPLP induced by increasing doses of histamine (top left panel), LTB$_4$ (top right panel) and LTD$_4$ (bottom panel). Each point is the mean ± SEM of 4–6 experiments. Contractions are expressed as percentages of those evoked by 5 μg (27 nmol) histamine. From Effects of gossypol on guinea-pig airway responsiveness: *in vitro* and *in vivo* studies. C.A. Touvay *et al* (1987), *Phytother. Res.*, **1**(2), 85–90, with permission.

In vivo *Bronchoconstriction in Anaesthetised Guinea-pigs: Effect of Various Natural Products*

Table 7.2 Effects of gossypol, NDGA and BN 52021 on PAF-acether-induced bronchoconstriction in guinea-pig. The substances were administered orally 1 h before challenge

	Dose (mg/kg)	No. of experiments	Maximal bronchoconstriction (%)	Variation (%)
Gossypol	–	4	83.7 ± 7.4	
	50.0	4	88.8 ± 4.4	+6.1
NGDA	–	4	83.7 ± 7.4	
	50.0	5	91.2 ± 1.8	+9.0
BN 52021	–	38	65.4 ± 3.1	
	10.0	9	28.1 ± 9.9	−57.0*

*p < 0.001 × difference from the control (Fisher's F test). NDGA = nordihydroguaiaretic acid; BN52021 = ginkgolide B; PAF = platelet activating factor. From Effects of gossypol on guinea-pig airway responsiveness: *in vitro* and *in vivo* studies. C.A. Touvay *et al* (1987), *Phytother. Res.*, 1(2), 85–90, with permission.

Table 7.3 Effects of gossypol, NDGA and BN 52021 on antigen-induced bronchoconstriction in sensitised guinea-pig. The substances were administered orally 1 h before challenge

	Dose (mg/kg)	No. of experiments	Maximal bronchoconstriction (%)	Variation (%)
Gossypol	–	13	49.3 ± 10.8	
	50.0	5	44.3 ± 22.1	−10.1
NGDA	–	13	49.3 ± 10.8	
	50.0	7	38.8 ± 16.4	−21.3
BN 52021	–	5	87.7 ± 5.9	
	20.0	5	42.2 ± 22.5	−51.9*

*p < 0.05 × difference from the control (Fisher's F test). NDGA = nordihydroguaiaretic acid; BN52021 = ginkgolide B; PAF = platelet activating factor. From Effects of gossypol on guinea-pig airway responsiveness: *in vitro* and *in vivo* studies. C.A. Touvay *et al* (1987), *Phytother. Res.*, 1(2), 85–90, with permission.

FURTHER READING

(*See also* references quoted in results section and those cited in review papers.)

General Physiology and Treatment of Asthma

Barnes PJ *et al* (1988). Inflammatory mediators and asthma. *Pharm. Rev.*, **40**, 49–84.
Barnes PJ (1989). Mechanism of disease: airway receptors. *Potgrad. Med. J.*, **65**, 532–542.
Barnes PJ and Liu SF (1995). Regulation of pulmonary vascular tone. *Pharm. Rev.*, **47**, 87–131.
Gould MK and Raffin TA (1995). Pharmacological management of acute and chronic bronchial asthma. In *Adv. Pharmacol.*, **32**, 169–204. Academic Press, London.
Marin MG (1986). Pharmacology of airway secretion. *Pharm. Rev.*, **38**, 273–289.
Marin MG (1994). Update: Pharmacology of airway secretion. *Pharm. Rev.*, **46**, 35–66.

PAF Antagonists: Reviews

Braquet P *et al* (1987). Perspectives in platelet activating factor research. *Pharmacol. Rev.*, **39**, 97–145.
Hosford D *et al* (1988). Natural antagonists of platelet activating factor. *Phytother. Res.*, **2**(1), 1–24.
Summers JB and Albert DH (1995). Platelet activating factor antagonists. In *Adv. Pharmacol.*, **32**, 67–168. Academic Press, London.

Leukotrienes and Lipoxygenase

Fitsimmons R *et al* (1986). Inhibition of human neutrophil 5-lipoxygenase activity by gingeridione, shogaol, capsaicin and related pungent compounds. *Prostaglandins Leukotrienes Med.*, **24**(2–3), 195–198.
Higgs GA and Moncada S (1985). Leukotrienes in disease: implications for drug development. *Drugs*, **30**, 1–5.
McMillan RM and Walker ERH (1992). Designing therapeutically effective 5-lipoxygenase inhibitors. *Trends Pharmacol. Sci.*, **13**, 323–330.
Piper PJ (1983). Pharmacology of leukotrienes. *Br. Med. Bull.*, **39**, 255–259.
Snyder DW and Fleisch JH (1989). Leukotriene antagonists as potential therapeutic agents. *Ann. Rev. Pharmacol. Toxicol.*, **29**, 123–143.
Wagner H (1989). Search for new plant constituents with potential antiphlogistic and antiallergic activity. *Planta medica*, **55**, 235–241.

Cytokines in Allergic Disease

Anderson GP and Coyle AJ (1994). T_H2 and 'T_22-like' cells in allergy and asthma: pharmacological perspectives. *Trends Pharmacol. Sci.*, **15**, 324–332.

Experimental

Burka JF (1985). Pharmacological modulation of responses of guinea-pig airways contracted with antigen and calcium ionophore A23187. *Br. J. Pharmacol.*, **85**, 411–425.

Constantine JW (1965). The spirally cut tracheal strip preparation. *J. Pharm. Pharmacol.*, **17**, 384–385.

Lulich KM *et al* (1979). The isolated lung strip and single open tracheal ring: a convenient combination for characterizing Schultz-Dale anaphylactic contractions in peripheral and central airways of the guinea-pig. *Clin. Exp. Pharmacol. Physiol*, **6**, 625–629.

Turner RA (1965). *Screening Methods in Pharmacology*, Academic Press, New York.

Antitussive Testing

Screening Methods in Pharmacology: as above.

Nicolis FB (1966). Evaluation of antitussive agents. In *International Encyclopedia of Pharmacology and Therapeutics. Section 6: Clinical Pharmacology*, Vol. 1, pp. 237–265, Pergamon Press, Oxford.

Packman EW and London SJ (1977). Clinical evaluation of antitussive agents. *Curr. Ther. Res.*, **21**(6), 855–866.

8
Anti-inflammatory and Analgesic Activity

The term inflammation covers a complex series of reparative and protective responses to tissue injury, whether caused by infection, auto-immune stimuli or mechanical injury.

Anti-inflammatory, analgesic and antipyretic activities have similar underlying mechanisms, but compounds differ in their profile of activity, e.g. corticosteroids are potent anti-inflammatory agents but are not analgesics; paracetamol and the narcotic analgesics have an analgesic effect with little anti-inflammatory activity; aspirin is both, like many other 'non-steroidal anti-inflammatory drugs' (NSAIDs). Anti-inflammatory drugs are used to treat disorders which lead to inflammation, pyrexia and pain of whatever cause, e.g. rheumatoid conditions, gout, dysmenorrhoea, neoplastic diseases and headache.

Most currently used anti-inflammatory agents, e.g. aspirin and the other NSAIDs, inhibit cyclo-oxygenase and therefore prostaglandin synthesis. These also have antipyretic activity, since prostaglandins are implicated in the mediation of fever. Free radical (FR) scavenging agents also play a role in inflammation, because liberation of FRs causes tissue damage during the inflammatory process. Flavonoids and other phenolics are thought to act by preventing the generation or action of free radicals. Quercetin has been shown to inhibit the action of lipoxygenase on arachidonic acid, the release of histamine from mast cells and basophils, and the generation of superoxide anion.

Testing for analgesic and anti-inflammatory action will be discussed separately. Platelet activating factor (PAF) involvement in inflammation is discussed in Chapter 7, 'The Respiratory System'.

TESTING FOR ANTI-INFLAMMATORY ACTIVITY

The main pathophysiological pathways for drug targeting at present are: arachidonic acid metabolism; the complement cascade; phagocytosis and other

131

cell functions; auto-immune processes; protein kinase C and other enzymes involved in second messenger systems.

The metabolism of arachidonic acid is shown schematically (Figure 6.13) in the section on platelets in Chapter 6; it is the target of most of the NSAIDs currently available. The complement system is an effector pathway of the non-specific humoral immune response and consists of a cascade of serum proteins, present in an inactive state until triggered by an antigen–antibody complex. The released peptide mediators exert multiple biological effects and are involved in diseases such as systemic lupus erythematosus and rheumatoid arthritis. Complement is therefore extremely important. There are specific *in vitro* complement assays based on the haemolysis of erythrocytes; these and the other more biochemical assays involving auto-immune and protein kinase C activity are beyond the scope of this book.

Inflammation is the product of these inter-related pathways and the response to infection, auto-immune or mechanical causes. Early inflammatory changes in damaged tissues are now known to involve the release of various biologically active materials from polymorphonuclear leukocytes (PMNs), lysosomal enzymes and others. The response is amplified by various chemotactic factors including 12-HETE (12-hydroxyeicosatetraenoic acid), a product of lipoxygenase activity on arachidonic acid. There is then a migration of macrophages into the inflamed tissues, which is a feature of the later stages of acute inflammation and of chronic inflammation. Once *in situ*, the macrophage becomes highly phagocytic and secretes an extensive range of products, including cyclo-oxygenase products and proteolytic enzymes. Platelets may contribute to the response by secreting similar products and also, in conjunction with thrombin, initiating complement activation. The vascular effects are primarily mediated by kinins, prostaglandins (PGs) and vaso-active amines (e.g. histamine, released by mast cells), which cause increased vascular permeability leading to plasma exudation. This is potentiated by PGE_2 and prostacyclin (PGI_2). These prostaglandins do, however, exhibit anti-inflammatory activity in some systems owing to their ability to elevate cAMP levels, which may inhibit PMN chemotaxis and macrophage migration. This anti-inflammatory action occurs mainly in the later, proliferative, rather than earlier, exudative, phase of inflammation.

It is therefore apparent that there are numerous sites at which drugs can act in anti-inflammation.

A simplified scheme of inter-related cells is shown in Figure 8.1 (Edwards, 1989), demonstrating the complexity of the inflammatory process and its interdependence on the immune system.

Animal Models: Anti-inflammatory Activity

Experimental inflammation in whole animals is the usual starting point for anti-inflammatory testing. These experiments are varied and widely used,

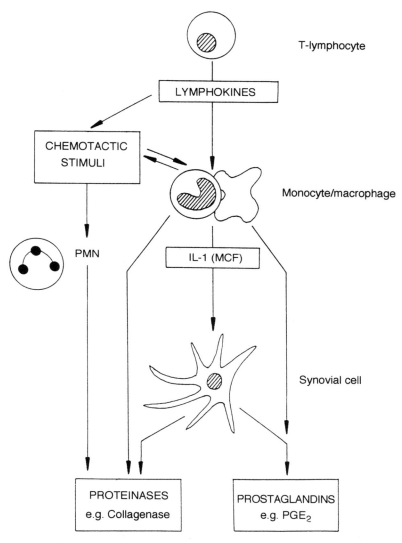

Figure 8.1 Scheme of cellular co-operation between T-lymphocytes, macrophages and synovial cells in the production of chronic inflammation.

especially the rat paw oedema test, and can be adapted in numerous ways using different inflammatory agents in an attempt to mimic pathological inflammation and arthritis. Some tests are relatively simple, such as those using erythema or oedema of the ear in mice or rats.

Cells involved in the inflammatory process, such as leucocytes, platelets and T-lymphocytes, can be employed, either intact or as homogenates which would, in effect, be enzyme assays. Platelet aggregation is discussed in Chapter 6. Measurement of the release of radio-labelled arachidonic acid, serotonin and other platelet and PMN constituents is also valuable.

Although various individual metabolic pathways are targeted for biochemical assays, a general *in vivo* test may involve all of these pathways. For example, in the rat paw oedema test, the carrageenan used to produce inflammation contains sulphated polysaccharides which are known to activate complement, and arachidonate metabolism is involved as shown by inhibition of inflammation in this model by the cyclo-oxygenase inhibitors indomethacin or aspirin. Most investigators, therefore, choose a range of tests in order to get an idea of the mechanism of action of an inhibitor.

The following experiments will be described; all can be used with both rats and mice: (a) mouse ear erythema or oedema; (b) rat paw oedema; (c) adjuvant arthritis in rat.

Enzyme assays will be discussed and these and other tests referenced.

Materials and Methods: Anti-inflammatory Activity (Figures 8.2–8.4: Tables 8.1–8.4)

Inflammation (Erythema and Oedema) of Rodent Ear

Animals

Mouse: albino (e.g. L.A.C.A.) adult, about 20 g body weight. Used for erythema and oedema measurement (it is essential to use albino mice for erythema experiments).
Rat: (e.g. Sprague–Dawley) adult, 50–70 g. Used for oedema measurements.

Irritants Used to produce Erythema or Oedema

These are usually dissolved in acetone to administer the dose in a volume of 5–10 μl:

Croton oil (1 mg/ear).
Isolated phorbol esters, e.g. TPA (tridecanoyl phorbol acetate), 12-DOPP (12-deoxyphorbolphenylacetate) and others (dose range 0.1–1.0 μg/ear).
Arachidonic acid (2 mg/ear, store in ice).
Ethyl phenyl proprionate (1 mg/ear).

Antagonists

Plant extracts under test for anti-inflammatory activity: dissolve in acetone, ethanol or 50% aqueous ethanol to give a suitable dose range (e.g. 0.1–5.0 mg/ear) in a volume of 10 μl.

Known antagonists if required—for comparison or when investigating mechanism of inflammation:
    Indomethacin: cyclo-oxygenase inhibitor (1 mg/ear).
    Quercetin: 5-lipoxygenase inhibitor (1 mg/ear).
    Hydrocortisone: corticosteroid (50 μU/ear).
    Mepyramine: H_1-receptor antagonist (1 mg/ear).
    Thioanisole: free radical scavenger (200 μg/ear).
    Propranolol: phospholipase inhibitor (1 mg/ear).
    Cyproheptadine: antihistamine, antiserotonin (1 mg/ear).

Method

Animals are divided into groups of 5–7 per dose. Plant extracts under test, or known anti-inflammatory agents, are applied to the pinna of the ear using a micropipette, about 15 min before applying the irritant (over the same area). Solvents may be tested for inflammatory or anti-inflammatory activity on the animals prior to the experiment.

Estimation of Inflammation

Erythema of mouse ear is the simplest experiment to perform as mice have translucent ears in which reddening is easily visible. It has the advantages that animals can be re-used several times and the course of inflammation can be followed over a period of time. Quantification is subjective but still valid: e.g. erythema can be described as pronounced, $+ +$; slight, $+$; or absent, 0.

Oedema is easier to quantify and is estimated by either punching a disc from the ear and weighing, or measuring the thickness using calipers. Measuring is only possible where oedema is pronounced; however, it has the advantage that it can also be measured at time intervals. *See* 'Treatment of Results' for examples.

Rat (or Mouse) Paw Oedema Test

Animals

Rats (Charles Forster, Wistar albino, etc.), 120–180 g body weight. Adrenalectomised rats may be used to eliminate involvement of the pituitary–adrenal axis.
Mice (Swiss albino) are occasionally used.

Pro-inflammatory Agents

These are drugs producing a measurable swelling (carrageenan is by far the most commonly used).

Carrageenan (0.1 ml of 1% w/v in 0.9% saline; 0.05 ml for mice).
Capsaicin (1–10 μg/kg in 10% ethanol/10% Tween 80/0.9% saline).
Dextrin (0.1 ml of 6% w/v in gum acacia 2% w/v).
Kaolin (0.1 ml, 5% suspension in 0.9% saline or 2% gum acacia).

Anti-inflammatory Agents

Plant extracts under test: suspend in 2–5% gum acacia or other suitable vehicle for oral or intraperitoneal administration. Usual dose 100 mg/kg of dry extract.
Known anti-inflammatory drugs if required for comparison or investigation of mechanism of action:
Indomethacin (cyclo-oxygenase inhibitor) 10 mg/kg p.o.
Diphenhydramine (antihistamine) 1 mg/kg i.p.
Methysergide (antiserotonin) 100 μg/kg i.p.
Corticosteroids.

Method

Animals are divided into groups of 6–8 per dose. The test extracts or antagonists are administered 1 h before the inflammatory agent if given orally, 30 min before if given i.p. Then inhibition of foot oedema is used as a measurement of anti-inflammatory activity. Oedema is produced by injecting the agonist into the plantar surface of the right hind-paw and the volume of the paw measured immediately and then at hourly intervals for up to about 5 h. It is measured by any suitable volume displacement method; instruments are available if required. Oedema is expressed as a mean increase in paw volume with respect to vehicle control; inhibition as a percentage increase or decrease in oedema volume. In mice it is usual to sacrifice the animals and sever both feet at the ankle, to measure the swollen paw relative to the untreated paw. For examples *see* 'Treatment of Results'.

Adjuvant Arthritis in Rat or Mouse

Animals

Rat: (Charles Forster), 120–180 g body weight.
Mice: (Swiss albino), 18–26 g.

Method

Groups of about five animals each per dose are used. Arthritis is induced by injecting the adjuvant, normally a suspension of killed *Mycobacterium tuberculosis*, 0.5% w/v in liquid paraffin, intradermally (i/d) into the hind paw

(0.05 ml for rats, 0.025 ml for mice). This is commercially available. Drug treatment for anti-inflammatory activity may start from the day before the adjuvant injection and continue as long as desired up to about 28 days, to give information about the development of arthritis and chronic drug treatment. Measurements are taken when swelling is established (usually by day 13) using the volume displacement methods outlined in the oedema test above. Test plant extracts (e.g. 100 mg/kg dry extract) suspended in gum acacia or other suitable vehicle are given either i.p. or p.o.

Enzyme Assays

These include cyclo-oxygenase and 5-lipoxygenase, the actions of which are shown in Figure 6.13.

Cyclo-oxygenase catalyses the metabolism of arachidonic acid to stable prostaglandins (PGE_2, PGD_2, $PGF_{2\alpha}$ and PGI_2) and thromboxane A_2. It may be obtained from sheep and pig seminal vesicles, rabbit or rat medulla of kidney. Whole cell systems such as platelets and polymorphonuclear leucocytes (PMNs) which involve cyclo-oxygenase are useful; platelets are discussed in Chapter 6 and PMNs are used in a similar manner.

5-Lipoxygenase is an enzyme complex which catalyses the oxygenation of higher unsaturated fatty acids in position C-5, followed by dehydration. This results in the formation of 5-HETE, 12-HETE and leukotrienes, especially LTB_4, which promote inflammatory processes and elicit allergic reactions such as asthma. It may be obtained from homogenates of intact pig or rabbit leucocytes, rat PMNs or human neutrophils.

Other Tests

The following tests have all been used successfully: cotton pellet granuloma; carrageenan-induced pleurisy; acetic acid-induced vascular permeability and turpentine-induced joint oedema.

For references *see* 'Further Reading'.

Treatment of Results

Erythema of Mouse Ear

Table 8.1 Inhibition produced by various antagonists of 12-DOPPA[a] (0.1 μg) induced erythema of mouse ears

Antagonist (1 mg per ear except where stated)	% Inhibition	
	2 h	4 h
Indomethacin	40	10
Phenol (50 μg)	25	25
Thioanisole (200 μg)	25	30
Aminopyrine	60	70
Desmethylimipramine	75	80
Imipramine	60	70
Promethazine	60	75
Trifluo perazine	75	95
Propranolol	70	70
p-Bromophenacyl bromide	15	50
Hydrocortisone (50 μg)	60	55

[a]12-DOPPA = 12-deoxyphorbol phenylacetate. From Inhibition of erythema induced by pro-inflammatory esters of 12-deoxyphorbol. Williamson EM and Evans F (1981), *Acta Pharmacol. Toxicol.*, **48**, 47–52, with permission.

Oedema of Mouse Ear

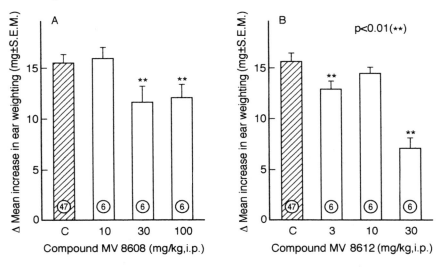

Figure 8.2   Effect of compounds MV8608 (A) and MV8612 (B) isolated from *M. velu-tina*, injected intraperitoneally 30 min before, on croton oil-induced ear oedema in mice. The number of animals in each group is represented at the base of each column, and the vertical bars indicate SEM. From Zanini, J. *et al* (1992) *Phytother. Res.*, **6**(1), 1–5.

Rat Paw Oedema—Carrageenan-induced

Table 8.2   Effect of topical application of IPA on carrageenan-induced paw oedema in rats

Treatment	n	Oedema volume (ml) (Hours after carrageenan injection			Inhibition (%) (Hours after carrageenan injection)		
		(2)	(3)	(5)	(2)	(3)	(5)
None	6	0.65 ± 0.06	0.62 ± 0.04	0.55 ± 0.06			
Acetone	12	0.62 ± 0.03^{ns}	0.64 ± 0.03^{ns}	0.52 ± 0.03^{ns}			
IPA							
0.4 mg/paw	8	0.48 ± 0.05^{a}	0.43 ± 0.04^{a}	0.31 ± 0.02^{b}	23	33	40
2 mg/paw	10	0.32 ± 0.03^{b}	0.37 ± 0.03^{b}	0.24 ± 0.03^{b}	48	42	54
4 mg/paw	7	0.33 ± 0.04^{b}	0.29 ± 0.04^{b}	0.23 ± 0.04^{b}	47	55	56
6 mg/paw	7	0.24 ± 0.03^{b}	0.22 ± 0.03^{b}	0.19 ± 0.02^{b}	61	66	63
Oxyphenbutazone							
2 mg/paw	7	0.38 ± 0.03^{b}	0.42 ± 0.02^{b}	0.36 ± 0.02^{a}	39	34	31

Results are means ±SEM.
[ns]Not significantly different ($p < 0.05$) from the control group receiving no treatment.
[a]Significantly different ($p < 0.05$) from the control group receiving acetone.
[b]Significantly different ($p < 0.001$) from the control group receiving acetone.
From Anti-inflammatory activity of *Ipomoea pes-caprae* (IPA). U. Pongprayoon *et al* (1991), *Phytother. Res.*, **5**(2), 63–66, with permission.

Rat Paw Oedema—Capsaicin-induced

Table 8.3 Effects of different inflammatory and anti-inflammatory agents on acute capsaicin-induced paw oedema in rats

Treatment and dose/kg body wt	Oedema vol. (ml) (mean increase in paw volume with respect to vehicle control $\pm$ SE)	Percentage increase (+) or decrease (−) of capsaicin-induced oedema vol. (compared to respective doses as control)
Capsaicin		
0.84 µg (s.c.)	0.13 ± 0.01	—
(extract) 21.00 µg (s.c.)	0.21 ± 0.02	—
84.00 µg (s.c.)	0.21 ± 0.02	—
Capsaicin		
0.84 µg (s.c.)	0.09 ± 0.01	—
(Sigma) 21.00 µg (s.c.)	0.23 ± 0.02	—
84.00 µg (s.c.)	0.23 ± 0.03	—
Carrageenan		
5 mg (s.c.)	0.29 ± 0.01	—
Capsaicin		
0.84 µg (s.c.) + carrageenan 5 mg (s.c. 60 min after)	0.46 ± 0.03[a]	+253.8
(extract) 21.00µg (s.c.) + carrageenan 5 mg (s.c. 60 min after)	0.78 ± 0.07[a]	+271.4
84.00µg (s.c.) + carrageenan 5 mg (s.c. 60 min after)	0.48 ± 0.03[a]	+128.6
Capsaicin		
0.84 µg (s.c.) + carrageenan 5 mg (s.c. 60 min after)	0.32 ± 0.03[a]	+255.6
(Sigma) 21.00µg (s.c.) + carrageenan 5 mg (s.c. 60 min after)	0.81 ± 0.06[a]	+252.2
84.00µg (s.c.) + carrageenan 5 mg (s.c. 60 min after)	0.54 ± 0.04[a]	+134.8
Capsaicin		
21.00 µg (s.c.) + indomethacin 10 mg (s.c. 60 min before)	0.18 ± 0.01	−14.3
(extract) 21.00µg (s.c.) + methysergide 0.1 mg (i.p. 30 min before)	0.19 ± 0.02	−9.5
21.00µg (s.c.) + diphenhydramine 1 mg (i.p. 30 min after)	0.09 ± 0.01[a]	−57.1

Control value for only vehicle (10% Tween 80 + 10% ethanol + normal saline) treated group was 0.04 ± 0.002. All s.c. injections were done with 0.05 ml of the drug into the plantar region of the right hind paw. Eight animals were used in each group.
[a]$p < 0.001$; Students' t-test with respect to capsaicin control. From De AK and Ghosh JJ (1990). Comparative studies on the involvement of histamine and substance P on the inflammatory response of capsaicin on the rat paw. *Phytother. Res.*, 4(1), 42–44, with permission.

Adjuvant Arthritis in Rat

Table 8.4 Effects of *Hymenocardia acida* extract and indomethacin on adjuvant arthritis in the rat

Treatment daily (p.o.)	Number of rats	Percentage inhibition of the primary lesion of arthritis at day 4 (right hind paw)	Percentage inhibition of secondary lesion of arthritis at day 20 (left hind paw)	Evidence of secondary lesions (ears and tail)
Saline				
0.2 ml	10	–	–	+ + + +
Indomethacin				
2 mg/kg	10	51.6 ± 3.7[a]	62.6 ± 5.1[a]	0
5 mg/kg	10	80.4 ± 4.8[a]	86.2 ± 6.2[a]	0
H. acida				
10 mg/kg	10	37.4 ± 3.5[a]	52.1 ± 4.6[a]	0
20 mg/kg	10	65.7 ± 5.1[a]	70.2 ± 6.3[a]	0

Figures given represent means of 10 paws ± SEM.
[a]Students' *t*-test significant differences between saline group and treated groups are denoted by $p < 0.001$; + + + + denotes the presence of severe secondary lesions at the tail and ears; 0 denotes absence of secondary lesions at the tail and ears. From Inhibition of adjuvant arthritis in the rat and pinnal inflammation in the mouse by an extract of *Hymenocardia acida*. Sackeyfio AC (1988), *Phytother. Res.*, 2(1), with permission.

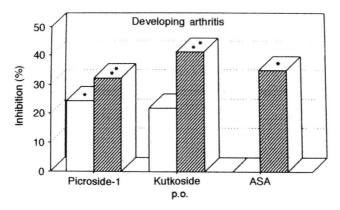

Figure 8.3 Effect of antiarthritic treatment (oral) with picroside-1 and kutkoside on development of left hind foot developing arthritis in rats following injection of 0.05 ml of adjuvant *Mycobacterium tuberculosis* (0.5% w/v) into left plantar aponeurosis. Each value is the mean of five animals. * $p < 0.05$, ** $p < 0.01$. □ 50 mg/kg; ▨ 100 mg/kg. From Anti-inflammatory activity of the iridoids kutkin, picroside-1 and kutkoside from *Picrorhiza kurroa*. Singh GB *et al* (1993), *Phytother. Res.*, 7(6), 402–407, with permission.

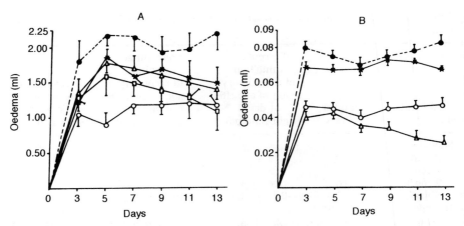

Figure 8.4 (A) Effect of anti-inflammatory treatment (oral) on development of left hind foot developing arthritis in rats following injection of 0.05 ml of adjuvant *Mycobacterium tuberculosis* (0.5% w/v) into left plantar aponeurosis. Each value represents the mean of five rats ± standard error (SE), control (●), kutkin 50 mg/kg (*), kutkin 100 mg/kg (△), kutkin 200 mg/kg (□), ASA 100 mg (○). (B) Effect of anti-inflammatory treatments (oral) on development of left hind foot developing arthritis in mice following injection of 0.025 ml of adjuvant *Mycobacterium tuberculosis* (0.5% w/v) into left plantar aponeurosis. Control (●), kutkin 50 mg/kg (*), kutkin 100 mg/kg (△), ASA 100 mg/kg (○). From Anti-inflammatory activity of the iridoids kutkin, picroside-1 and kutkoside from *Picrorhiza kurroa*. Singh GB *et al* (1993), *Phytother. Res.*, 7(6), 402–407, with permission.

TESTING FOR ANALGESIC AND ANTIPYRETIC ACTIVITY

The previous methods described for anti-inflammatory testing will also show positive for those analgesics, such as aspirin, which act on peripheral cyclo-oxygenase pathways. They will not, however, reveal opiate-type analgesic action and compounds such as paracetamol (acetaminophen), which work on cyclo-oxygenase in the brain only.

Testing for analgesic activity is the type of classical pharmacology well covered in the literature (*see* 'Further Reading'). However, as it is so important it will be reviewed briefly and the most commonly used tests described.

Pain is difficult to define and measure, and all methods are therefore subjective or semi-quantitative, but good results can be obtained if these limitations are taken into account.

The sensation of pain is initiated in peripheral pain receptors (nociceptors) and its purpose is to draw attention to tissue damage. Unfortunately, the sensation tends to hang on after this purpose has been accomplished. The impulses generated are conveyed to the CNS by afferent pain fibres and thence to the dorsal horn, spinal cord, reticular formation and thalamus and finally the cerebral cortex. Thus many parts of the brain are involved in the perception of pain. This is complicated further by the fact that emotional experience can dominate the CNS and intensify or block pain.

Analgesics can therefore work in several ways, and it is for this reason that they are often used in combination, mainly a narcotic-type with an anti-inflammatory or paracetamol.

Narcotic analgesics work by mimicking natural neurotransmitter peptides known as endorphins and enkephalins and others. There are several opioid receptors known, the main CNS receptors being the δ (delta), κ (kappa) and μ (mu), with others including the σ (sigma) and ε (epsilon) receptors.

Morphine, the oldest and one of the most widely used of the opiate analgesics, is known to act primarily at μ receptors. Naloxone antagonises drug action at μ, δ and κ receptors.

In order to test for analgesic activity it is obviously necessary to induce pain in the subject and then modify the response to, or perception of, this pain (c.f. anaesthesia, where the *passage* of pain impulses to the CNS is inhibited). This causes some difficulty: in animal experiments it is assumed that the animal responds to a pain stimulus in a similar manner to that which a human would, which cannot be proved. However, correlations have confirmed the usefulness of these tests. Experimentally induced pain in human volunteers, e.g. ischaemic pain, would seem to be ideal but in fact shows greater difficulties owing to the subjective nature of pain and the psychological component which is missing from experimentally induced pain. In a clinical situation where double-blind trials are used, up to 20% of patients will find pain relief from placebos, therefore animal experiments are still important.

The main methods of inducing pain experimentally are: *thermal*, e.g. hot-plate test, tail-flick test; *mechanical*, e.g. tail-clip method; *chemical*, e.g. writhing (squirming) test; *electrical*, e.g. stimulation of tooth-pulp in man; *ischaemic*, e.g. application of a tourniquet to the arm, in man. The estimation of pain is either by *quantal response*, in which the percentage of animals responding to a fixed stimulus, e.g. heat, is determined; or a *threshold response*, in which the stimulus, e.g. pressure/heat, is increased until each animal responds.

Inhibition of prostaglandin synthesis is also related to the control of body temperature during fever (pyrexia). Leucocytes release interleukin-1, which acts on the thermoregulatory (TR) centre in the hypothalamus, causing an elevation in temperature. There is also an associated rise in prostaglandin (PG) levels in the brain, which also act on the TR centre. Aspirin, paracetamol and other cyclo-oxygenase inhibitors prevent the rise in PG levels and hence the rise in temperature. Aspirin also inhibits the effects of interleukin-1 on temperature elevation.

Testing for antipyretic activity can be carried out by testing against experimentally induced hyperthermia, as well as the previous anti-inflammatory tests which give a good indication of cyclo-oxygenase inhibition.

Animal Models: Analgesic and Antipyretic Activity

Previously, tests were divided into those distinguishing narcotic from non-narcotic analgesics. However, this is no longer useful, especially when screening, and it is preferable to carry out several different tests and use observation and specific antagonists to elucidate the mechanism of action.

The following tests will be described:

1. Writhing (squirming) test.
2. Hot-plate test.
3. Tail-flick test.

Use of Naloxone Antagonism: Antipyretic Testing

The writhing test uses the observation that an i.p. injection of a mildly irritant substance causes the mouse to writhe or squirm. Both narcotic and non-narcotic analgesics will prevent writhing, but the test can be modified to include an estimation of capillary leakage, as measured by inclusion of a plasm-bound dye. This will be inhibited by a non-narcotic, but not a narcotic, analgesic (*see* 'Further Reading'). There are limitations to the writhing test, e.g. drugs other than analgesics will inhibit writhing; these include antihistamines, sympathomimetics and parasympathomimetics, CNS stimulants such as amphetamine, and adrenergic blockers. However, the test is sensitive: a very low dose of morphine is detectable, lower than that needed for the hot-plate or tail-flick test. During this and other tests it is useful to look for the Traub effect—an erect tail—which is indicative of a narcotic effect; this is how pethidine was discovered. Not all animals show the writhing reaction and it is necessary to do a preliminary test to make sure the animals to be used will do so.

The hot-plate test uses a heat stimulus and is simple and efficient. Mice or rats are used, and the hot-plate maintained at a temperature a little too hot for comfort. The time taken for the animal to either lick its paws or jump out of the cylinder surrounding the hot-plate is a measure of analgesic effect.

The tail-flick test depends on the reaction of a rat or mouse to a painful stimulus on the tail, which is a violent jerking action. The stimulus is usually heat, either a water bath or a radiant heat source. Again, it is advisable to check before the test that the animals to be used will respond in this way.

Naloxone antagonism can be tested with any of the experiments described to see if narcotic activity is involved. Naloxone is administered prior to the plant extract and subsequent testing: any analgesic effects due to opiate-type compounds will be abolished, since naloxone is an opiate antagonist devoid of agonist activity, blocking all opiate receptor types so far identified. *See* use in conjunction with tail-flick test in 'Treatment of Results'.

Antipyretic testing may be carried out by testing the effect of plant extracts on hyperthermia in rats induced by a pyrogen such as an injection of yeast.

Materials and Methods: Analgesic and Antipyretic Activity (Figures 8.5–8.6; Tables 8.5–8.8)

Writhing Test

Animals

Mice, adult, in groups of 6–12 per dose.

Agonists (i.e. irritants to cause writhing reaction)

Acetic acid, 1–3%, 0.2 ml, injected i.p., or 300 mg/kg.
Benzoquinone, 0.2 mg/ml, 0.25 ml, injected i.p.

Antagonists

Test plant extracts: 0.5 g–1 g/kg, in water or propylene glycol (or other), administered p.o., s.c. or i.p.
Control: same volume of vehicle by same route.
Known antagonists if required for comparison:
Indomethacin 10 mg/kg p.o.: cyclo-oxygenase inhibitor.
Phenacetin 500 mg/kg i.p.: cyclo-oxygenase inhibitor.
Morphine 5 mg/kg s.c.: narcotic analgesic.

Method

Test substances and known antagonists are given 15–30 min prior to the i.p. injection of irritant. Acetic acid is the most commonly used. After about 5 min the number of writhes in a set period, such as 30 min, are counted and noted with a suitable counter.

Hot-plate Test

Animals

Mice, adult, less frequently rats, in groups of 6 or more per dose.

Hot-plate

Maintained at a temperature of 50–58 °C, with a constraining cylinder to prevent animals jumping off too easily.

Antagonists

Plant extracts: 0.5–1.0 g/kg either p.o. or i.p.
Control: same volume of vehicle by same route.

Known antagonist if required: morphine 5 mg/kg s.c.

Note: non-narcotic antagonists do not work well in this test.

Method

Animals are dropped gently onto the hot-plate and the reaction time measured. The only responses considered are licking of paws or jumping off the hot-plate and out of the cylinder; other types of behaviour are ignored. The reaction time (mean for the group) can then be plotted against dose. Alternatively, reaction time can be measured for animals kept on the hot-plate for a maximum of 30 sec, and those animals in which the reaction time is increased to at least twice that of the controls are taken to show significant analgesia. The ED_{50} can then be calculated as the dose at which 50% of the animals show analgesic effects.

Tail-flick Test

Animals

Rats, (less frequently mice), adult, in groups of six or more per dose

Antagonists

Test plant extracts, 0.5 g–1 g/kg, in water or propylene glycol (or other), administered p.o., s.c. or i.p.
Control: same volume of vehicle by same route
Known antagonists if required for comparison:
Phenacetin 500 mg/kg i.p.: cyclo-oxygenase inhibitor.
Morphine 5 mg/kg s.c.: narcotic analgesic.
Fentanyl 0.2 mg/kg s.c.: narcotic analgesic.

Method

Test substances and known antagonists are given 15–30 min before the animal is subjected to the pain stimulus. The tail is immersed, either totally or just the terminal 2 cm, in a hot water bath at a constant temperature (55–58 °C). Alternatively, a radiant heat source is focused on the tail; equipment is available specifically to do this. The time until reaction (a jerk of the tail) is recorded, and the test repeated at suitable time intervals from about 15 min after injection of test substances. If required, the ED_{50} can be defined as the dose causing 50% of the animals to respond later than a given time interval (e.g. 5 sec).

Naloxone Antagonism

Dose of naloxone: 0.5–10 mg/kg, i.p., administered 15 min prior to test substance (or morphine, 5 mg/kg s.c., for comparison), followed by chosen method of analgesic testing, methods as above.

Antipyrexia Testing

Animals

Rats, adult, in groups of at least six per dose.

Pyrogen

Brewer's yeast, 15% suspension, 1 ml/kg administered s.c.

Antagonists

Test plant extracts: 0.5 g–1 g/kg, administered p.o., s.c. or i.p.
Control: same volume of vehicle by same route.
Known antipyretic if required for comparison:
Aspirin 100 mg/kg i.p.

Method

Yeast extract is administered to the animals and after 10–15 h the rectal temperature determined. This is the initial pyretic temperature. Test extracts, at doses similar to those used for analgesic testing, are then administered by a suitable route and the rectal temperature recorded at hourly intervals. The difference between the mean temperature of the control group and the test group can then be calculated.

Treatment of Results

Testing for Analgesic Activity

Writing Test

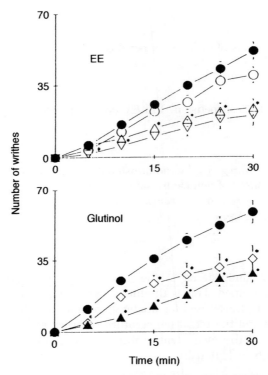

Figure 8.5 Accumulative number of writhings induced by intraperitoneal injection of 0.8% acetic acid (0.1 ml/10 g) in mice pretreated orally with either the vehicle (●, control), the ethanolic extract of *Scoparia dulcis* (○, EE: 0.25 g/kg; △, 0.5 g/kg; ▽, 1 g/kg) (upper figure), glutinol (◇, 30 mg/kg) or indomethacin (▲, 10 mg/kg) (lower figure). Symbols and vertical bars are means ±SE of at least 10 animals. * Different from control ($p < 0.05$). From Analgesic and anti-inflammatory properties of *Scoparia dulcis*. L. extracts and glutinol in rodents. S M de Farias Freire *et al* (1993), *Phytother. Res.*, **7**(6), 408–414, with permission.

Table 8.5 Effects of terpenoids and reference drugs on acetic acid-induced writhing syndrome in mice. Drugs were administered 30 min before intraperitoneal injection of 3% acetic acid 0.2 ml

Compound	Dose (mg/kg i.p.)	No. of writhings (mean $\pm$ SE)	Inhibition vs. controls (%)
Controls		18.8 $\pm$ 1.15	
Disidein triacetate	5	19.0 $\pm$ 1.11	1.2
	10	18.4 $\pm$ 1.32	2.1
2-Polyprenyl benzoquinols	5	16.5 $\pm$ 0.37	12.2
	10	11.0 $\pm$ 1.15[a]	41.5
2-Tetraprenyl benzoquinol	5	9.2 $\pm$ 0.86[b]	51.1
	10	5.6 $\pm$ 0.81[b]	70.2
4-Hydroxy-3-tetraprenyl benzoic acid	5	15.6 $\pm$ 0.74	17.0
	10	13.8 $\pm$ 0.86[a]	26.6
Avarol	5	13.0 $\pm$ 0.7[a]	30.8
	10	8.6 $\pm$ 0.5[b]	54.2
Furospongin-1	5	18.0 $\pm$ 0.94	4.2
	10	13.0 $\pm$ 0.7[a]	30.8
Phenacetin	500	11.4 $\pm$ 0.9[a]	39.4
Morphine	5 (s.c.)	3.0 $\pm$ 0.7[b]	84.0

[a] $p < 0.05$; [b] $p < 0.001$ vs. controls; Student's t-test. From Pharmacological studies on terpenoids from marine sponges: analgesic and muscle relaxant effects. R de Pasquale *et al* (1991), *Phytother. Res.*, **5**(2), 49–53, with permission.

Hot-plate Test

Table 8.6 Hot-plate test in rat. Effects of terpenoids and reference drugs on reaction time, 30 and 90 min after drug administration

Compound	Dose (mg/kg i.p.)	Reaction time (sec) mean $\pm$ SE	
		30 sec	90 sec
Controls		85.0 $\pm$ 4.5	79.2 $\pm$ 6.6
Disidein triacetate	5	91.2 $\pm$ 5.0	86.6 $\pm$ 6.7
	10	93.1 $\pm$ 2.8	85.7 $\pm$ 5.3
2-Polyprenyl benzoquinols	5	101.1 $\pm$ 5.0[a]	98.3 $\pm$ 8.0
	10	98.4 $\pm$ 4.0[a]	100.0 $\pm$ 4.0[a]
2-Tetraprenyl benzoquinol	5	93.4 $\pm$ 4.8	89.5 $\pm$ 4.5
	10	104.5 $\pm$ 5.0[a]	120.0 $\pm$ 5.0[b]
4-Hydroxy-3-tetraprenyl benzoic acid	5	111.3 $\pm$ 4.7[b]	109.9 $\pm$ 6.7[a]
	10	110.0 $\pm$ 7.0[a]	108.6 $\pm$ 7.0[a]
Avarol	5	98.2 $\pm$ 2.8	101.7 $\pm$ 3.7
	10	119.0 $\pm$ 1.3[a]	111.1 $\pm$ 3.0[b]
Furospongin-1	5	95.4 $\pm$ 2.1	90.4 $\pm$ 2.3
	10	91.3 $\pm$ 3.7	98.5 $\pm$ 6.8
Phenacetin	50	97.6 $\pm$ 1.9	109.0 $\pm$ 16.0
Morphine	5 (s.c.)	155.0 $\pm$ 16.0[b]	136.0 $\pm$ 14.0[b]

[a] $p < 0.05$; [b] $p < 0.001$ vs controls; Student's t-test. From Pharmacological studies on terpenoids from marine sponges: analgesic and muscle relaxant effects. R de Pasquale *et al* (1991), *Phytother. Res.*, **5**(2), 49–53, with permission.

Tail-flick Test in Rat

Table 8.7 Tail-flick test in rat. Effects of terpenoids and reference drugs on reaction time measured 30 and 90 min after drug administration

Compound	Dose (mg/kg i.p.)	Reaction time (sec) mean $\pm$ SE	
		30 sec	90 sec
Controls		13.9 ± 0.6	13.8 ± 1.0
Disidein triacetate	5	15.6 ± 1.0	15.3 ± 0.6
	10	19.0 ± 1.8^a	18.4 ± 1.4^a
2-Polyprenyl benzoquinols	5	20.5 ± 0.8^b	20.7 ± 0.9^a
	10	24.1 ± 1.2^b	21.4 ± 1.2^a
2-Tetraprenyl benzoquinol	5	18.4 ± 0.9^a	17.5 ± 0.6^a
	10	21.2 ± 1.0^b	20.0 ± 1.0^a
4-Hydroxy-3-tetraprenyl benzoic acid	5	26.8 ± 1.5^b	26.4 ± 1.2^a
	10	27.7 ± 2.0^b	24.6 ± 1.9^a
Avarol	5	17.6 ± 0.56^a	18.4 ± 0.7^a
	10	22.7 ± 1.5^b	22.8 ± 1.0^b
Furospongin-1	5	12.1 ± 0.49	12.5 ± 0.5
	10	12.3 ± 0.64	12.8 ± 0.28
Phenacetin	500	15.4 ± 0.6	15.9 ± 0.4^a
Morphine	5 (s.c.)	57.1 ± 2.6^b	46.9 ± 2.6^b

[a] $p < 0.05$; [b] $p < 0.001$ vs controls; Student's t-test. From Pharmacological studies on terpenoids from marine sponges: analgesic and muscle relaxant effects. R de Pasquale *et al* (1991), *Phytother. Res.*, **5**(2), 49–53, with permission.

Tail-flick Test Using Naloxone Antagonism

Table 8.8 Antagonism of naloxone against analgesic effect of terpenoids and morphine on nociceptive response induced in rat by tail-flick

Compound	Reaction time (sec) mean $\pm$ SE	
	30 sec	90 sec
Controls	13.0 ± 0.3	13.1 ± 0.5
Naloxone	14.2 ± 0.4	13.4 ± 0.3
2-Polyprenyl benzoquinols	24.1 ± 1.2^b	21.4 ± 1.2^b
Naloxone + 2-polyprenyl benzoquinols	13.1 ± 0.9^c	14.6 ± 0.5^c
2-Tetraprenyl benzoquinol	21.2 ± 1.0^b	20.0 ± 1.0^b
Naloxone + 2-Tetraprenyl benzoquinol	12.7 ± 0.6^c	13.5 ± 0.4^c
4-Hydroxy-3-tetraprenyl benzoic ac.	27.7 ± 2.0^b	24.6 ± 1.9^b
Naloxone + tetraprenyl benzoic ac.	27.4 ± 1.3^b	20.5 ± 1.3^b
Avarol	22.7 ± 1.5^b	22.1 ± 1.0^b
Naloxone + avarol	19.8 ± 1.2^b	18.8 ± 1.1^a
Morphine	57.1 ± 2.6^b	46.9 ± 2.6^b
Naloxone + morphine	12.8 ± 0.7^c	12.9 ± 0.4^c

Terpenoids (10 mg/kg, i.p.); Naloxone (1 mg/kg, i.p.); morphine (5 mg/kg, s.c.). [a] $p < 0.05$; [b] $p < 0.001$ vs. controls; [c] $p < 0.001$ test drugs + naloxone vs. test drugs; Student's t-test. Pharmacological studies on terpenoids from marine sponges: analgesic and muscle relaxant effects. R de Pasquale *et al* (1991), *Phytother. Res.*, **5**(2), 49–53, with permission.

Testing for Antipyretic Activity

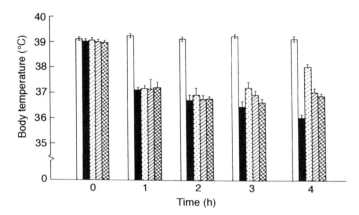

Figure 8.6 Effect of *S. cuminii* seed extract and aspirin on yeast-induced pyrexia in rats (temperatures expressed as mean $\pm$ SE, $n = 6$). (□) Vehicle control; (■) acetyl-salicylic acid (100 mg/kg); (▨) extract (50 mg/kg); (▨) extract (100 mg/kg); and (▨) extract (200 mg/kg). Control vs. extract-treated at 1, 2 and 3 h $p < 0.001$, and at 4 h $p < 0.01$. Control vs. acetylsalicylic acid, at all times $p < 0.001$. From Anti-inflammatory and related actions of *Syzygium cuminii* seed extract. A.K. Nag Chandhuri *et al* (1990), *Phytother. Res.*, 4(1), 5–14, with permission.

FURTHER READING

Anti-inflammatory Testing

(*See also* references quoted in 'Treatment of Results')

Methods

Various Tests

Bouclier M *et al* (1990). Experimental models in skin pharmacology. *Pharm. Rev.*, **42**, 127–154.

Chaudhuri A *et al* (1990). Anti-inflammatory and related actions of *Syzygium cuminii* seed extract. *Phytother. Res.*, **4**(1), 5–10.

Hostettmann K (Ed) (1991). *Methods in Biochemistry, Vol. 6. Assays for Bioactivity*, Academic Press, London, and references therein.

Singh GB *et al* (1993). Anti-inflammatory activity of the iridoids kutkin, picroside-1 and kutkoside from *Picrorhiza kurroa*. *Phytother. Res.*, **7**(6), 402–407.

Mouse Ear Erythema

Evans FJ and Schmidt RJ (1979). An assay procedure for comparative irritancy testing of esters in the tigliane and daphnane series. *Inflammation*, **3**, 215–223.

Lewis, DA (1989) Assessment of antiinflammatory activity. In *Antiinflammatory Drugs from Plant and Marine Sources. Agents Actions Supplement. Vol 27.* Birkhäuser Verlag, Basel.

Williamson EM and Evans FJ (1981). Inhibition of erythema induced by pro-inflammatory esters of 12-deoxyphorbol. *Acta Pharmacol. Toxicol.*, **48**, 47–52.

Zanini J et al (1992). Action of compounds from *Mandevilla velutina* on croton oil-induced ear oedema in mice. A comparative study with steroidal and non-steroidal anti-inflammatory drugs. *Phytother. Res.*, **6**(1), 1–5.

Mouse Ear Erythema/Platelet Aggregation

Edwards MC et al (1983). New phorbol and deoxyphorbol esters: isolation and relative potencies in inducing platelet aggregation and erythema of skin. *Acta Pharmacol. Toxicol.*, **53**, 177–187.

Enzyme Assays

Flower RJ (1978). Prostaglandins and related compounds. In *Handbook of Experimental Pharmacology*, Vol. 50(1), pp. 394–422, Eds. JR Vane and SH Ferreira. Springer-Verlag, Heidelberg.

Glaser KB (1995). Regulation of phospholipase A_2 enzymes: selective inhibitors and their pharmacological potential. In *Adv. Pharmacol.* **32**, 31–66. Academic Press, London.

Nugteren DH and Christ-Hazelhof E (1980). Chemical and enzymic conversion of the prostaglandin endoperoxide PGH_2. In *Advances in Prostaglandin and Thromboxane Research*, pp. 129–137 Vol. 6, Eds. B. Samuellson et al. Raven Press, New York.

Anti-inflammatory Activity of Flavonoids

Cody V, Middleton E and Harborne (Eds) (1986). *Plant Flavonoids in Biology and Medicine*, Vols. I and II, Alan R. Liss, New York.

Rat Paw Oedema Testing

Vinegar R et al (1987). Pathway to carrageenan-induced inflammation in the hind limb of the rat. *Fed. Proc.*, **46**, 118–126.

Winter CA et al (1962). Carrageenan-induced oedema in hind paw of rat as an assay for anti-inflammatory drugs. *Proc. Soc. Biol. Med.*, **111**, 544–547.

Winter CA (1964). Anti-inflammatory testing methods: comparative evaluation of indomethacin and other agents. In *International Symposium on Non-steroidal Anti-inflammatory Drugs*, pp. 190–202, Eds. S Garantini and M Dukes. Exerpta Medica Foundation, Amsterdam.

Adjuvant Arthritis Test

Katz L and Piliero SJ (1969). A study of adjuvant-induced polyarthritis in the rat with special reference to associated immunological phenomena. *Ann. N.Y. Acad. Sci.*, **147**, 515–536.
Newbould BB (1963) Chemotherapy of arthritis induced in rats by mycobacterial adjuvant. *Br. J. Pharmacol.*, **21**, 127–136.

Capillary Permeability and Writhing

Whittle BA (1964). The use of changes in capillary permeability in mice to distinguish between narcotic and non-narcotic analgesics. *Br. J. Pharmacol.*, **22**, 246–253.
Williams TJ (1981). Mediation of increased vascular permeability after complement activation. *J. Exp. Med.*, **153**, 136–153.

Cotton Pellet Granuloma Test

Winter CA and Porter CC (1957). Effects of alteration in side chain upon antiinflammatory and liver glycogen activities of hydrocortisone esters. *J. Am. Pharm. Ass.*, **46**, 515–519.

Turpentine-induced Joint Oedema

Hanson JM *et al* (1974). Anti-inflammatory property of 401 (MCD-Peptide), a peptide from the venom of the bee *Apis mellifera*. *Br. J. Pharmacol.*, **50**, 383–392.

Carrageenan-induced Pleurisy

Meacock SCR and Kitchen EA (1979). Effects of non-steroidal anti-inflammatory drug benoxaprofen on leucocyte migration. *J. Pharm. Pharmacol.*, **31**, 366–370.

Analgesic and Antipyretic Testing

General Pharmacology of Analgesics

Gilman LS Ed. *et al* (1990). *Goodman and Gilman's Pharmacological Basis of Therapeutics*, 8th Edn, Pergamon Press, New York.
Martindale (1993). *Extra Pharmacopoeia*, 30th Edn. Pharmaceutical Press, London.
Neal MJ (1992). *Medical Pharmacology at a Glance*, 2nd Edn, Blackwell Scientific, Oxford.
Okpako DT (1991). *Principles of Pharmacology: A Tropical Approach*, Cambridge University Press, Cambridge.
Pleuvry JP (1991). Opioid receptors and their ligands: natural and unnatural. *Br. J. Anaesth.*, **66**, 370–380.

General Screening Methods

Edvinsson L (1995). Experimental headache models in animals and humans. *Trends Pharm. Sci.*, **16**, 5–9.
Turner A (1965). *Screening Methods in Pharmacology*, Academic Press, New York and references therein.

Other Methods

Writing Test

Berkenkopf JW and Weichman BM (1988). Production of prostacyclin in mice following i.p. injection of acetic acid, phenylbenzoquinone and zymosan: its role in the writhing response. *Prostaglandins*, **36**, 693–709.

Okun R *et al* (1963). The effects of aggregation, electric shock and adrenergic blocking drugs on the inhibition of the "writhing syndrome". *J. Pharmacol. Exp. Ther.*, **139**, 107–109.

Siegmund EA *et al* (1957). A method for evaluating both non-narcotic and narcotic analgesics. *Proc. Soc. Exp. Biol.*, **95**, 729–731.

Hot-plate Test

Eddy NB and Leimbach D (1953). Synthetic analgesics II. Dithienylbutenyl and dithienylbutylamines. *J. Pharmacol. Exp. Ther.*, **107**, 385–393.

MacDonald AD *et al* (1946). Analgesic actions of pethidine derivatives and related compounds. *Br. J. Pharmacol.* **1**, 4–14.

Tail-flick Test

Davies OL *et al* (1946). Method for evaluation of analgesic activity using rats. *Br. J. Pharmacol.*, **1**, 255–264.

Janssen PAJ *et al* (1963). The inhibitory effect of fentanyl and other morphine-like analgesics on the warm water induced tail withdrawal reflex in rats. *Arzneimittelforsch. Drug Res.*, **6**, 502–507.

Ramabadran K *et al* (1989). Tail immersion test for the evaluation of a nociceptive reaction in mice. Methodological considerations. *J. Pharmacol. Meth.*, **21**, 21–31.

9

Diabetes Mellitus

Diabetes mellitus is a common, chronic metabolic disorder involving carbohydrate, fat and protein metabolism and characterised by glycosuria and hyperglycaemia. Two forms of diabetes mellitus are usually described: insulin dependent diabetes mellitus (IDDM), or type 1; and non-insulin dependent diabetes mellitus (NIDDM), or type 2.

Type 1, IDDM, has been called juvenile-onset diabetes since it often presents in children or young adults. It is due to a deficiency in insulin synthesis and secretion from the β-cells in the islets of Langerhans found in the pancreas. This may be due to an auto-immune disorder (anti-islet cell antibodies are sometimes found) or caused by a virus such as Coxsackie B_4 or mumps. Treatment of type 1 diabetes is with insulin and diet; there is no real place for oral hypoglycaemics.

Type 2, NIDDM, usually occurs in elderly and/or overweight people, and has been called maturity-onset, or mild obese, diabetes. It is caused by a relative deficiency, or diminished effectiveness, of insulin and is much more common than IDDM. NIDDM may also be secondary to other diseases such as pancreatitis, or a result of insulin antagonism caused, for example, by Cushing's syndrome or steroid therapy. A genetic component appears to be involved when NIDDM occurs in young people. Other causes of impaired glucose tolerance include pregnancy and thyrotoxicosis, and the gradual development of relative insulin deficiency is part of the normal ageing process. In NIDDM, treatment is to regularise blood sugar levels by the use of hypoglycaemics, α-glucosidase inhibitors, sugar-restricted diets, and complex carbohydrate preparations which delay the absorption of glucose from the gut.

TESTING FOR HYPOGLYCAEMIC ACTIVITY

The discovery of the oral synthetic hypoglycaemics was a major advance in the treatment of diabetes. The sulphonylureas, for example, act by augmenting insulin secretion, and the biguanides only work in the presence of residual insulin.

There are many plant extracts with proven hypoglycaemic activity such as jambul, *Syzygium cumini*; fenugreek, *Trigonella foenum-graecum*, garlic, *Allium sativum*; onion, *Allium cepa*; ginseng, *Panax ginseng*; guava, *Psidium guava*; bitter gourd, *Momordica charantia*, and others such as *Nelumbo nucifera*, *Opuntia ficus indica* and *Vernonia amygdalina*. No isolated compounds from these plants have yet been developed for clinical use, but all are used in traditional or alternative medicine to treat diabetes mellitus. Many of those which are dietary components have been tested in human diabetics and found to be hypoglycaemic.

The oral sulphonylureas in clinical use are hypoglycaemic in normal (i.e. non-diabetic) animals, and successful screening programmes have been carried out using normal rabbits and rats; however, nowadays diabetic animals are usually employed as well. Diabetes may be induced by administering a diabetogenic toxin such as alloxan or streptozocin, and the hypoglycaemia or glucose tolerance produced by a plant extract compared to that of a standard such as tolbutamide. Recently, a new class of drugs, the α-glucosidase inhibitors such as acarbose, have been made available. These retard digestion and absorption of carbohydrates (*see* Clissold and Edwards (1988)).

Clinical testing in human non-insulin dependent diabetics involves measuring blood glucose levels after administering the test extract, after either overnight fasting or preloading with glucose.

Animal Models: Hypoglycaemic Activity

Normoglycaemic animals

Normal animals are often used for testing potential oral hypoglycaemics; this is still a valid screening method and is often used in addition to diabetic animal models. The comparison may give some information regarding mechanism of action. Hyperglycaemic agents can also be detected at the same time. A single plant extract (e.g. garlic) may contain compounds having both actions, which can complicate results.

Animals can be fasted prior to treatment or given a glucose loading and a standard glucose-tolerance test (GTT) performed. Usually rabbits or male rats are used but the methods apply to humans, provided that there are no toxicity problems.

Insulin levels can be measured after treatment with a hypoglycaemic agent to see whether stimulation of insulin secretion from the pancreatic β-cells has occurred; and comparisons between insulin levels found in fasting and glucose-loaded animals show whether it is the plant extract or the glucose itself which is acting as an insulin secretagogue.

It may also be useful to measure liver glycogen or triacylglycerol levels, to give some idea of whether utilisation or absorption of glucose has been affected, since the hypoglycaemic activity of some medicinal plants may involve mechanisms of action other than pancreatic stimulation. For details, *see* Day *et al* (1993); Nyarko *et al* (1993).

In vitro glucose uptake studies can be carried out on rat hemidiaphragms to see if insulin effects are being enhanced: *see* Huralikuppi *et al* (1991).

Alloxan-diabetic Animals

Alloxan is a urea derivative which causes selective necrosis of the pancreatic islet β-cells. It is used to produce experimental diabetes in animals such as rabbits, rats, mice and dogs. It is possible to produce different grades of severity of the disease by varying the dose of alloxan used; these may be classified by measuring fasting blood sugar (FBS) levels: e.g. in rabbits moderate diabetes has been defined as an FBS level of 180–250 mg/ml, and severe diabetes as an FBS level of above 250 mg/ml (Huralikuppi *et al* (1991)).

The *severe* diabetes produced by alloxan results in blood sugar levels equivalent to a total pancreatectomy, hence sulphonylureas such as tolbutamide, which act mainly by stimulating insulin release from β-cells, show only a small hypoglycaemic effect in this instance. Therefore a test plant extract producing a significant hypoglycaemia (in a *severely* alloxan-diabetic animal) must be operating through a different mechanism (e.g. somatostatin, gastro-intestinal hormones, corticosteroids, prostaglandins), or affecting glycogen or glucose metabolism: *see* Hikino *et al* (1989).

For testing drugs for use in NIDDM, and for studies on glucose tolerance, moderately diabetic animals are more often used.

Streptozocin-diabetic Animals

Streptozocin (also called streptozotocin) is a cytotoxic nitrosoureido glucopyranose derivative isolated from fermentations of *Streptomyces achromogenes*; it produces diabetes in animals. The induction of diabetes takes some time, and experiments carried out at suitable time intervals after administration of streptozocin will give additional information into mechanism of action. During the earlier part of this induction period, hypoglycaemic action of plant extracts may be due to stimulation of the residual activity of β-cells; however, after induction of the diabetic state, hypoglycaemic activity will be due to some other mechanism, since in this model of diabetes insulin activity is thought to be negligible. An example of non-insulin mediated hypoglycaemia is that due to the hypoglycans, e.g. lithosperman A, from *Lithospermum erythrorizan* and moran A from *Morus alba*, which affect glucose metabolism and cause a rise in blood sugar levels immediately after administration, followed by a fall. *See* Hikino *et al* (1985, 1986); Niyonzima *et al* (1993).

Streptozocin is considered to be carcinogenic and should be treated with caution.

Genetically Obese Diabetic and Non-diabetic Mice

Mutant strains of obese diabetic mice are available, such as the C57BL/KsJ-*db/db*, their non-diabetic (but still obese) equivalent, C57BL/6J-*ob/ob*; and the lean diabetic and non-diabetic littermates of both of these. It is therefore possible to test for effects of plant extracts on blood glucose as well as body weight, insulin production and insulin resistance in this model.

The pathogenesis of diabetes in the *db/db* mouse develops in stages, the first lasting up to about 3 months and characterised by obesity, normal pancreatic insulin and hyperinsulinaemia, suggesting that the islet β-cells can meet the apparent increased insulin demand; next there is marked obesity with extreme hyperinsulinaemia and decreased pancreatic insulin, indicating that insulin demand is not being met; and finally pancreatic and blood levels of insulin fall as the β-cells lose their function and become necrotised, and there is a loss of weight before death at 6–8 months.

The *db/db* mouse responds to insulin at a very young age but continued insulin treatment does not prevent the development of diabetes or insulin resistance. This indicates that the diabetes associated with this mutant strain is non-insulin dependent (type 2). Compounds such as the sulphonylureas which stimulate insulin release from the β-cells therefore have an effect during the early stages of this model of diabetes but not at the later stages when the β-cells have been destroyed.

Mice of the obese (but non-diabetic strain *ob/ob* have persistent hyperinsulinaemia compared with their lean controls, but the levels of hyperglycaemia are variable between different colonies.

Comparisons between hypoglycaemia induced in diabetic and non-diabetic mice by a plant extract over a period of time will show whether the effect is by preventing the rise in blood sugar associated with this model of diabetes, or a hypoglycaemic effect *per se*. *See* Addy *et al* (1992).

Materials and Methods: Antidiabetic Testing (Figures 9.1–9.6: Tables 9.1–9.2)

Induction of Diabetes

Alloxan-induced Diabetes in Rabbits, Rats and Mice

For all animals a single dose of alloxan, 140–180 mg/kg (usually 150 mg/kg) is administered as a 5% w/v in distilled water, injected i.v. into the marginal ear vein of the rabbit or i.p. in the case of mice and rats. A rest period of 7 days for rabbits and 2 days for rats and mice is allowed, during which the animals have free access to food and water.

Streptozocin-induced Diabetes in Rats and Mice

A single dose of streptozocin (SZ) in sterile citrate buffer (e.g. pH 4.5, 0.1M) may be used: mice 150 mg/kg; rats 80 mg/kg, administered i.p. Diabetes develops gradually and may be assessed after a few days, usually 4 days for mice and 7 days for rats. Experiments may then be carried out at increasing time intervals as the disease progresses, if desired.

To demonstrate that diabetes has been produced, blood sugar levels are determined; these should be high, e.g. around 200–500 mg/100 ml, although this will depend on the degree of diabetes required (normal blood glucose levels in rabbits are around 110–130 mg/100 ml, in rats and mice 70–80 mg/100 ml).

Sometimes diabetic animals are maintained on insulin until needed, e.g. in the rat, 2 units protamine zinc insulin may be given daily, the last injection at least 24 hours before the experiment.

Genetically Obese Diabetic Mice

If the *db/db* mouse model is used, animals are normally kept on a diet of the test extract for a time-dependent study over a fairly long period (e.g. 90 days), for reasons already discussed. The effect on body weight is determined at the same time, and results of these studies are more complicated (*see* 'Treatment of Results').

Glucose Tolerance Testing (GTT)

GTT is a standard procedure, involving monitoring blood glucose levels over a period of time following a glucose load. It is used in the diagnosis of diabetes, and in experiments such as these where a test extract is being assessed for hypoglycaemic activity. Animals are fasted overnight prior to the procedure, then a glucose load (1 g/kg) given either with or without the test hypoglycaemic agent. Blood may be taken from the marginal ear vein of the rabbit and the tail vein in the mouse and rat, or elsewhere. 200 μl is sufficient for most analyses. For time scales and doses, *see* examples in 'Treatment of Results'.

Measurement of Blood Glucose

Since blood glucose measurement is carried out routinely in homes, hospitals and laboratories all over the world by both patients and professionals, there are many readily available techniques, proprietary kits, reagent testing strips and portable machines. Any of these giving reasonable accuracy are suitable. Glucose autoanalysers are widely used, e.g. from Beckman, Ames, Roche. If these are not available, there are two classical methods, using *ortho*-toluidine or glucose oxidase, which are described in detail in Bauer (1982).

Measurement of Blood Insulin Levels

It may be useful to determine whether there is a rise in circulating insulin levels after glucose loading, and during treatment with plant extract. The usual methods are RIA (radio-immunoassay) and ELISA (enzyme-linked immunosorbent assay), available as kits, e.g. from Boehringer, Mannheim; Novo Biolabs, Copenhagen; Pharmacia, Brussels; and others. For examples, *see* 'Treatment of Results'.

Clinical Testing in Human Subjects

Human non-insulin dependent diabetic volunteers can be used to test non-toxic plant extracts and in principle are the same as the animal experiments described. However, the two examples shown in 'Treatment of Results' also demonstrate different mechanisms: there is the true hypoglycaemic effect of the *Opuntia ficus indica* extract (Figure 9.4), and the blunted insulin release and post-prandial glycaemic profiles resulting from ingestion of guar gum, *Cyamopsis tetragonolobus* (Figure 9.5). Guar gum, which is a viscous fibre, has been shown to act by delaying gastric emptying. In the *Opuntia* experiment this was not the case, since patients did not receive a glucose load before the plant extract.

Treatment of Results

Glucose Tolerance Tests in Normoglycaemic Mice With and Without Plant Extract

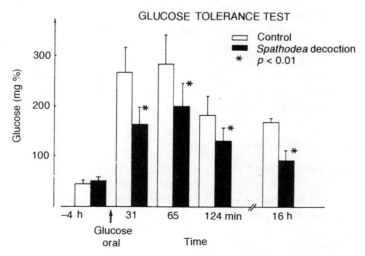

Figure 9.1   Results of GTT. Each column represents the mean + SEM of at least six different mice. From Hypoglycaemic activity of *Spathodea Campanulata* stem bark decoction in mice. Niyonzima G *et al* (1993), *Phytother. Res.*, 7(1), 64–67, with permission.

Table 9.1   Glucose tolerance tests (GTT) on mice

Time (h)	Protocol used in Figure 9.1
−4	Injection of decoction (i.p.) 8 g bark powder/kg
0	Tail puncture (blood glucose) Oral glucose 150 mg/mice
0.5–2	Tail puncture (blood glucose)
16	Cardiac puncture (plasma insulin)

From Hypoglycaemic activity of *Spathodea campanulata* stem bark decoction in mice. Niyonzima G *et al* (1993), *Phytother. Res.*, 7(1), 64–67, with permission.

Hypoglycaemic Effect of a Plant Extract Compared with Chlorpropamide

Table 9.2   Effect of aqueous extract (800 mg/kg) and chlorpropamide (400 mg/kg) on alloxan-induced diabetic rabbits (three animals per experiment)

Treatment	Mean percentage deviation from initial blood glucose concentration $\pm$ SEM (Hours after administration)			
	(2)	(4)	(8)	(12)
Control	$+0.58 \pm 0.52$	$+1.33 \pm 1.20$	-2.49 ± 2.24	-1.80 ± 1.62
Aqueous extract	-10.96 ± 4.42	$-29.58^c \pm 38.13$	$-37.80^c \pm 48.70$	$+3.30 \pm 4.26$
Chlorpropamide	$-21.19^a \pm 20.59$	$-39.05^c \pm 37.96$	$-27.62^b \pm 27.33$	$+0.62 \pm 0.63$

[a] $p < 0.025$; [b] $p < 0.0025$; [c] $p < 0.005$ compared with control, Student's *t*-test. From Hypoglycaemic activity of *Anthocleista vogellii* aqueous extract in rodents. Abuh FY *et al* (1990), *Phytother. Res.*, 4(1), 20–24, with permission.

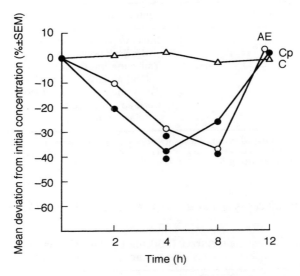

Figure 9.2   Effect of aqueous extract (AE, 800 mg/kg) and chlorpropamide (Cp. 200 mg/kg) on alloxan-induced diabetic rabbits. Control (C) * $p < 0.005$. From Hypoglycaemic activity of *Anthocleista vogellii* aqueous extract in rodents. Abuh FY *et al* (1990), *Phytother. Res.*, **4**(1), 20–24, with permission.

Effect of a Plant Extract on Streptozocin-induced Diabetes in the Rat

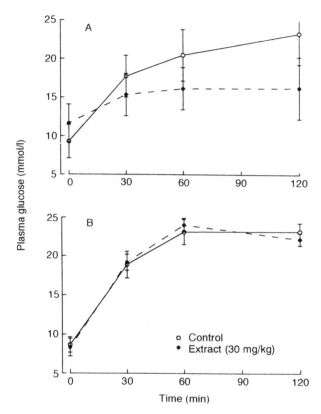

Figure 9.3 Effect of freeze-dried extract of *Indigofera arrecta* (30 mg/kg body weight) on oral glucose challenge in streptozotocin-induced diabetic rats, (A) 7 days and (B) 17 days after induction of diabetes. Vertical bars indicate standard errors of the means, $n = 4$. From The basis for the antihyperglycaemic activity of *Indigofera arrecta* in the rat. Nyarko AK *et al* (1993), *Phytother. Res.*, 7(1), 1–4, with permission.

Clinical Studies Using Human Non-insulin Dependent Diabetic Subjects

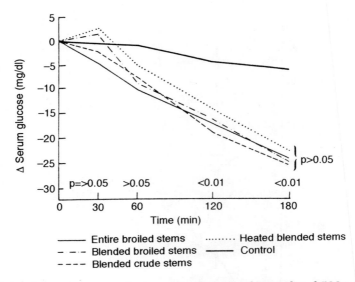

Figure 9.4 Change in serum glucose concentrations after intake of 500 g of *Opuntia ficus indica* stems prepared in various ways compared to 400 ml water (control). From Hypoglycemic effect of *Opuntia ficus indica* in non-insulin dependent diabetes mellitus patients. Frati A *et al* (1990), *Phytother. Res.*, **4**(5), 195–197, with permission.

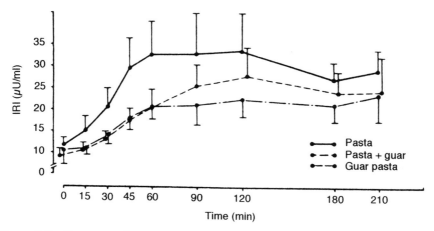

Figure 9.5 Plasma insulin concentrations (mean ± SEM) before three test-meals and for 210 min after each meal. From Guar-enriched pasta and guar gum in the dietary treatment of type 2 diabetes. Briani G *et al* (1990), *Phytother. Res.*, **1**(4), 177–179, with permission.

Measurement of Plasma Insulin Levels in Normoglycaemic Rats after Glucose Loading and Treatment with Plant Extract

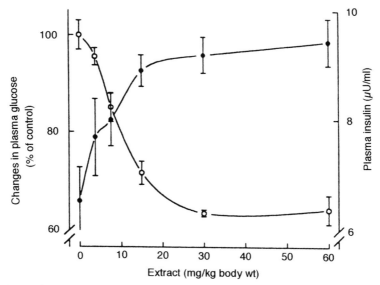

Figure 9.6 Changes in plasma glucose and insulin levels of normoglycaemic rats in response to different concentrations of freeze-dried extract of *Indigofera arrecta*, 60 min after oral glucose challenge. Vertical bars indicate standard errors of the means, $n = 4$. ○, Changes in glucose; ●, plasma insulin. From The basis for the antihyperglycaemic activity of *Indigofera arrecta* in the rat. Nyarko AK *et al* (1993), *Phytother. Res.*, **7**(1), 1–4, with permission

FURTHER READING

General: Includes Experimental Details

Ellenberg M and Rifkin H (Eds) (1983). *Diabetes Mellitus: Theory and Practice*, 3rd Edn. Medical Examination Publishing Co. Inc., New York.

Oral Hypoglycaemics and Dietary Control of Diabetes

Clissold SP and Edwards C (1988). Acarbose: A preliminary review of its pharmacodynamic and pharmacokinetic properties, and therapeutic potential. *Drugs*, **35**, 214–243.

Hall SE et al (1980). The effect of bran on glucose kinetics and plasma insulin in non-insulin dependent diabetes mellitus. *Diabetes Care*, **3**, 520–525.

Hikino H et al (1989). Mechanisms of hypoglycaemic activity of Ganoderman B: a glycan of *Ganoderma lucida* fruit bodies. *Planta Medica*, **55**, 423–428.

Huralikuppi JC et al (1991). Antidiabetic effect of *Nelumbo nucifera* extract: Part 2. *Phytother. Res.*, **5**, 217–223.

Jenkins DJA et al (1980). Diabetic diets: high carbohydrate combines with high fibre. *Am. J. Clin. Nutr.*, **33**, 1729–1733.

Nakahara K et al (1994). Inhibition of post-prandial hyperglycaemia by Oolong tea extract. *Phytother. Res.*, **8**(7), 433–435.

Pederson O et al (1982). Increased insulin receptor binding to monocytes from insulin-dependent diabetic patients after a low fat, high starch, high fibre diet. *Diabetes Care*, **5**, 284–291.

Perl M. (1988). The biochemical basis of the hypoglycemic effects of some plant extracts. In Cracker, LE and Simon, JE (Eds), *Herbs, Spices and Medicinal Plants*, Vol. 3, Oryx Press, Phoenix, Arizona.

Srivastava Y et al (1993). Antidiabetic and adaptogenic properties of *Momordica charantia* extract: an experimental and clinical evaluation. *Phytother. Res.*, **7**(4), 285–289.

Reviews

Bailey CJ et al (1989). Drugs inducing hypoglycaemia. *Pharmacol. Ther.*, **42**, 361–384.

Gerich JE (1989). Oral hypoglycaemic agents. *New Engl. J. Med.*, **321**, 1231–1245.

Rerup CC (1980). Drugs producing diabetes through damage of the insulin secreting cells. *Pharm. Rev.*, **22**, 485–517.

Watkins JB and Sanders RA (1995). Diabetes mellitus-induced alterations of hepatobiliary function. *Pharm. Rev.*, **47**(1), 1–23.

Methods

Streptozocin-induced Diabetes

Hikino H et al (1986). Isolation and hypoglycemic activity of Moran A, a glycoprotein of *Morus alba* roots. *Planta Medica*, **52**, 64–65.

Konno C et al (1985). Isolation and hypoglycemic activity of Lithospermans A, B and C, glycans of *Lithosperm erythorhizon* roots. *Planta Medica*, **51**, 157–160.

Like AA and Rossini AA (1976). Streptozotocin-induced pancreatic insulitis; a new model of diabetes mellitus. *Science*, **193**, 415–417.

Niyonzima G *et al* (1993). Hypoglycaemic activity of *Spathodea campanulata* stembark decoction in mice. *Phytother. Res.*, **7**(1), 64–67.

Nomikos IN *et al* (1989). Involvement of O_2 radicals in 'autoimmune' diabetes. *Immunol. Cell Biol.*, **67**, 85–87.

Papaccio G *et al* (1991). Superoxide dismutase in low dose streptozocin-treated mice. *Int. J. Pancreatol.*, **10**, 253–260.

Tomlinson KC *et al* (1992). Functional consequences of streptozotocin-induced diabetes mellitus, with particular reference to the cardiovascular system. *Pharmacol. Rev.*, **44**(1), 103–150.

Genetically Diabetic Mice

Addy M *et al* (1992). *Indigofera arrecta* prevents the development of hyperglycemia in the *db/db* mouse. *Phytother. Res.*, **6**(1), 25–28.

Bray GA and Yoy DA (1971). Genetically transmitted obesity in rodents. *Physiol. Rev.*, **511**, 598–646.

Genuth SM (1969). Hyperinsulinism in mice with genetically determined obesity. *Endocrinology*, **84**, 386–391.

Normoglycaemic Animals

Neef H *et al* (1995). Hypoglycaemic activity of selected European plants. *Phytother. Res.*, **9**(1), 45–48.

Measurement of Blood Glucose

Bauer J (1982). *Clinical Laboratory Methods*, 9th Edn, C.V. Mosby, St Louis.

Dubowski KM (1962). An *o*-toluidine method for body fluid glucose determination. *Clin. Chem.*, **8**, 215–221.

Related Experimental Methods

Asayama K *et al* (1984). Chemiluminescence as an index of drug-induced free radical production in pancreatic islets. *Diabetes*, **33**, 160–163.

Murat JC and Serfaty A (1974). Simple enzymatic determination of polysaccharide (glycogen) content of animal tissue. *Clin. Chem.*, **20**, 1576–1577.

Perl M and Hikino H (1989). Effect of some hypoglycemic glycans on glucose uptake and glucose metabolism by inverted intestinal fragments. *Phytother. Res.*, **3**(5), 433–435.

Suzuki Y and Hikino H (1989). Mechanisms of hypoglycemic activity of panaxans A and B, glycans of *Panax ginseng* roots: effects on the key enzymes of glucose metabolism in the liver of mice. *Phytother. Res.*, **3**(1), 15–19.

10

The Nervous System

THE EFFECT OF PLANT EXTRACTS ON THE PERIPHERAL NERVOUS SYSTEM

Introduction

The major components of the peripheral nervous system are the *autonomic* (consisting of the *parasympathetic* and *sympathetic* divisions) and the *motor nerve–skeletal muscle* nervous systems (Figure 10.1 and Table 10.1). The autonomic nervous system (ANS) is the branch of the nervous system that can function even when it is uncoupled from the central nervous system (CNS). Stimulation of the sympathetic branch of the ANS in the intact animal, as in fright, gives rise to increased rate of beating and force of contraction in the heart, contraction of some blood vessels (those supplying the gut) and dilatation of others (those supplying skeletal muscles and the heart). Stimulation of the parasympathetic branch (e.g. the vagus nerve) slows the heart, contracts most non-vascular smooth muscles and increases glandular secretions. The transmitter for sympathetic nerves is noradrenaline (NA), and for parasympathetic nerves, acetylcholine (ACH). Ongoing research suggests that adenosine triphosphate (ATP) and nitrous oxide (NO) may be neurotransmitters or co-transmitters in some peripheral nerves named *purinergic* and *nitriergic*.

Preganglionic nerves of the ANS leave different sections of the spinal cord to synapse with cell bodies called ganglia, from where *postganglionic* nerves radiate to innervate different organs. The most important transmitter of impulses from the *pre-* to the *postganglionic* nerve is ACH in both divisions of the ANS. Common targets for drug action in the ANS are: (a) neurotransmitter receptors; (b) release mechanisms; (c) synthesis; (d) degradation/uptake mechanisms; or (e) storage of neurotransmitter. Plant extracts may, therefore, be examined for effects on these mechanisms.

The motor nerve–skeletal muscle junction is the target for the action of muscle relaxants acting peripherally. The motor nerve neurotransmitter is

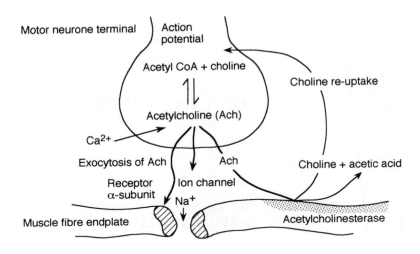

EVENTS AT THE NEUROMUSCULAR JUNCTION

SITE OF DRUG ACTION

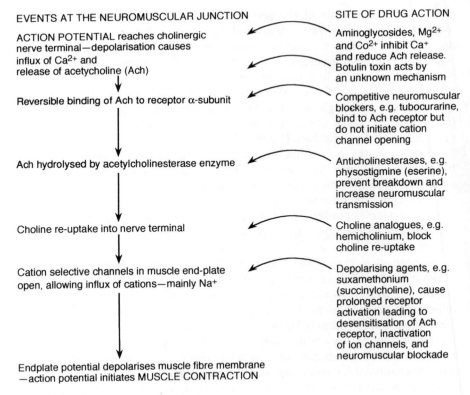

EVENTS AT THE NEUROMUSCULAR JUNCTION	SITE OF DRUG ACTION
ACTION POTENTIAL reaches cholinergic nerve terminal—depolarisation causes influx of Ca^{2+} and release of acetycholine (Ach)	Aminoglycosides, Mg^{2+} and Co^{2+} inhibit Ca^+ and reduce Ach release. Botulin toxin acts by an unknown mechanism
Reversible binding of Ach to receptor α-subunit	Competitive neuromuscular blockers, e.g. tubocurarine, bind to Ach receptor but do not initiate cation channel opening
Ach hydrolysed by acetylcholinesterase enzyme	Anticholinesterases, e.g. physostigmine (eserine), prevent breakdown and increase neuromuscular transmission
Choline re-uptake into nerve terminal	Choline analogues, e.g. hemicholinium, block choline re-uptake
Cation selective channels in muscle end-plate open, allowing influx of cations—mainly Na^+	Depolarising agents, e.g. suxamethonium (succinylcholine), cause prolonged receptor activation leading to desensitisation of Ach receptor, inactivation of ion channels, and neuromuscular blockade

Endplate potential depolarises muscle fibre membrane
—action potential initiates MUSCLE CONTRACTION

Figure 10.1   The neuromuscular junction.

Table 10.1 The autonomic nervous system

BODY SYSTEM Organ	Adrenergic receptor	Effect of sympathetic stimulation	Effect of parasympathetic stimulation (all receptors cholinergic, muscarinic)
DIGESTIVE			
Gut wall	β_1, β_2, α	Decrease in tone and motility	Increase in tone and motility
Sphincters			
Salivary gland	α	Secretion of thick saliva	Copious secretion of watery saliva
Pancreas			Increase in endo- and exocrine secretion
CARDIOVASCULAR			
Heart	β_1	Increase in rate and force of beat	Reduction in rate and force of beat
Blood vessels	α	Vasoconstriction	
	β_2	Vasodilatation	
RESPIRATORY			
Airways	β_2	Bronchodilatation	Bronchoconstriction Bronchosecretion
LIVER	α, β_2	Release of glucose into bloodstream	
EYE	α	Dilatation of pupil	Constriction of pupil, tear secretion
BLADDER	α, β	Relaxation of detrusor, excitation of sphincter	Detrusor excited. sphincter relaxed (for micturition)
UTERUS	α, β_2	Contraction or relaxation depends on hormonal state	
SKIN	α	Piloerection	
PENIS	α	Ejaculation	Venous sphincters contract to maintain erection
ADRENAL MEDULLA ⟶		RELEASE OF ADRENALINE	

ACH. Well known drugs of plant origin with actions at the motor end-plate are tubocurarine (ACH receptor antagonist) and physostigmine (acetylcholinesterase inhibitor) The action of physostigmine is not restricted to the motor nerve end-plate; it is observed at all sites where ACH is released. The ANS and the neuromuscular junction (NMJ) are summarised briefly here; full details are available from any of the pharmacology textbooks listed.

IN VITRO MODELS FOR INVESTIGATIONS OF THE PERIPHERAL NERVOUS SYSTEM (FIGURES 10.2–10.5)

Guinea-pig Ileum (GPI)

The isolated segment of the guinea-pig ileum has a number of applications and advantages for the study of agents acting on the ANS:

1. When properly set up, the GPI is devoid of large spontaneous contractions that may obscure agonist-induced contractions.
2. Embedded in the wall of the GPI, between the longitudinal and circular muscle layers, is a network of intrinsic parasympathetic nerves (the Auerbach and Meissner's plexuses); there are also non-adrenergic non-cholinergic (NANC) nerves. When suitably mounted between platinum electrodes, these nerves can be transmurally stimulated to cause muscle contraction. Nerve-mediated as well as directly stimulated contractions can thus be studied in the GPI.
3. The Auerbach's plexus contains parasympathetic ganglion bodies which can be stimulated by nicotinic receptor agonists, e.g. nicotine; agents with ganglion stimulating or blocking properties can thus be investigated also on the isolated GPI.
4. A wide range of substances can cause contraction of the GPI; therefore, agents with specific agonist/antagonist action at histamine H_1-receptors, ACH muscarinic/nicotinic receptors and 5-HT (serotonin) receptors can be studied in this preparation qualitatively and quantitatively. Useful information can also be derived about the mode of action of antispasmodic agents acting non-specifically at sites beyond agonist–receptor combination, e.g. on cation channels.
5. The preparation is easy to make, robust, capable of reproducible responses for many hours, and does not require investment in high technology. In the absence of electronic devices, a frontal writing lever and smoked paper can be used effectively to record contractions of the GPI.

SETTING UP THE GPI

Many investigators take it for granted that the GPI is a simple preparation to make, and therefore do not take due care in setting it up, and end up with avoidable failure. Whatever the form of recording to be used, every care should be taken to ensure that the gut muscle is not damaged during the process of setting it up. A guinea-pig is deprived of food, but with access to water, overnight. The animal is killed by a stunning blow to the head and immediately cutting the neck vessels. When the abdomen is cut open, the ileum can be found attached to the back of the caecum. Avoiding the first 5 cm next to the caecum, a convenient length is freed from its mesenteric attachment and transferred to a Petri dish containing suitable physiological salt solution (PSS). The contents of the lumen are flushed out with the PSS contained in a 10 ml pipette, the tip of which is placed over one end of the ileum, with the minimum pressure (do not blow into the pipette!).

A 2 cm length of ileum is set up in a jacketed organ bath of 5–10 ml capacity. The PSS, maintained at an approximately constant temperature (32–37 °C) is aerated with air or a gas mixture of 95% oxygen and 5% CO_2. One end of the ileum is attached to a tissue holder at the base of the organ bath and the other to a recording device—isotonic or isometric. The length of the ileum should be about 1.5–2.0 cm. In general, for a given set of calibrations of the recording device, the size of response is proportional to the length of gut; the length should be such that a maximum response can be recorded, and submaximal responses of reasonable size can be reproduced. This comes from practice and experience. The ligatures should be threaded through the muscle from the inside of the lumen, on the side opposite to the mesenteric attachment (as in Figure 10.2). The lumen of the gut is thus open, allowing debris to be voided throughout the experiment.

SOME USES OF THE ISOLATED GPI

Test of Plant Extract for Parasympathomimetic or Parasympatholytic Activity

Reference has already been made to the use of this preparation for testing for non-specific antispasmodic activity (*see* Chapter 4). The object here is to describe the uses of the GPI to test for specific activity on the parasympathetic nervous system. Agents which promote the activity of the parasympathetic nervous system, such as anticholinesterases, e.g. physostigmine (from *Physostigma venenosum*), or act directly as ACH receptor agonists, such as arecoline (from *Areca catechu*) and pilocarpine (from *Pilocarpus jaborandi*) are called *parasympathomimetic* or *cholingergic*. Agents that do the opposite, that is, prevent the action of ACH, such as atropine (from *Atropa belladona*) or its release, such as morphine (*Papaver somniferum*) are *parasympatholytic* or *anticholinergic*. Both types of substances are used in medicine to increase or decrease intestinal motility. Anticholinergic drugs are widely used to control diarrhoea and in the treatment of Parkinson's disease, travel sickness, peptic ulcer and anticholinesterase poisoning.

Procedure

The GPI is set up as described above and allowed to equilibrate with regular changes of bath fluid. Graded doses of ACH are used to stimulate the ileum, each dose being allowed a contact time of 10–15 sec. Doses of the extract are then added.

Parasympathetic (Cholinergic) Mechanism of Action   The extract may produce a contraction; a cholinergic mechanism should be suspected if the contractions, like those of ACH, persist in the presence of a histamine H_1-receptor antagonist (e.g. mepyramine, about 10^{-8}M), which will block equivalent responses to histamine. A cholinergic mechanism is likely if atropine (10^{-8}M) blocks the response to ACH and the extract, but there is not an equivalent effect on response to histamine. Atropine will not distinguish a purely *muscarinic receptor* effect from a *nicotinic* action at the ganglia; stimulation of ganglion receptors leads to the release of ACH at the end of the postganglionic parasympathetic nerve. The released ACH will then produce a contraction by action at muscarinic receptors on the muscle cell membrane. This is blocked by atropine. A ganglion-blocking drug such as hexamethonium can be used to make this distinction. A nicotinic receptor mechanism should be suspected if, in the presence of coniine (from *Conium maculatum*) or hexamethonium (10^{-4}M), contractions caused by nicotine (from *Nicotiana tabacum*) ($10\,\mu g/ml$ or more is needed in the isolated GPI) as well as those of the extract, are blocked, but not those of ACH.

The contraction caused by the extract may be due to actions at other surface receptors such as 5-HT, for which several specific binding ligands are now available, or histamine; a specific anti-5-HT_3-receptor activity would be very valuable, as such agents are known to have anti-emetic or antinausea profiles. When testing a plant extract for specific receptor activity, a known agonist or antagonist at the suspected receptor must be used as a control; confidence can be increased further if specificity at another receptor system is established by the use of specific agonist/antagonist under the experimental conditions in operation.

Parasympatholytic (Anticholinergic) Mechanism of Action   The extract causes no contraction but inhibits contractions caused by other agonists. Specific anticholinergic activity is suspected if moderate doses of extract block ACH contractions, but not contractions to histamine and 5-HT. If the preliminary observations suggest a muscarinic type of block, confirmation can be sought by the use of dose–response curves, from which the nature of the antagonism (competitive or non-competitive) and relative antagonistic potency (pA_2 values) can be obtained (*see* Chapter 2).

Block of Intramural Nerves   A nicotinic receptor type of block can be established as described above by using nicotine and hexamethonium in comparison with the extract. A number of substances are known to inhibit parasympathetic nerve function by preventing the release of ACH from the cholinergic nerve endings. An example of a drug that acts by this mechanism is morphine. The procedure for testing for this kind of activity involves mounting the GPI as described above between a pair of platinum electrodes. Instead of using the whole ileum, a longitudinal muscle strip can be used; the longitudinal muscle is stripped away from the underlying circular muscle layer, such that the Auerbach's plexus remains attached intact to the longitudinal muscle strip (Paton & Zar, 1968). When stimulated at maximum voltage, with pulse widths of 0.1–0.5 ms, the muscle gives repeated contractions. The response is proportional to the frequency of stimulation. The contraction is due to ACH released from parasympathetic nerve endings and is inhibited by drugs like morphine (pre-junctional) and atropine (post-junctional). When released, ACH is rapidly hydrolysed by acetylcholinesterase to yield choline and acetic acid. Some of the choline is transported back into the neurone for use in the synthesis of new ACH. Therefore, for this procedure to yield reproducible results for many hours, it is advisable to include choline in the PSS. Another precaution is not to use pulse widths higher than about 0.5 ms. High pulse widths cause muscle contraction by direct depolarisation. To check that the contractions of the transmurally stimulated GPI is indirectly mediated, add tetrodotoxin (10–100 ng/ml), when the twitches should be quickly extinguished. With pulse widths of about 0.5 ms, stimulation at high frequencies (10/s) may give tetrodotoxin-sensitive responses that are resistant to atropine due to stimulation of so-called non-adrenergic non-cholinergic (NANC) nerves. The physiological role of such nerves is not yet ascertained.

The Isolated Rabbit Jejunum and Adrenergic Mechanisms (Figures 10.3–10.4)

The isolated rabbit jejunum (IRJ) has a number of advantages as a method for the study of adrenergic mechanisms:

1. When properly set up, the muscle exhibits rhythmic regular contractions of impressive amplitude that are sustained for many hours.
2. The muscle can be set up with the periarterial nerve threaded through an electrode, so that the sympathetic nerve supply to the muscle can be stimulated. The response of the muscle to sympathetic nerve stimulation is inhibition of spontaneous contractions and reduction in resting tone. This response is due to noradrenaline released from noradrenergic nerve endings; preparations made from reserpine-treated rabbits respond poorly or respond with a contraction to periarterial nerve stimulation.
3. Since the muscle has inherent tone and contracts spontaneously, noradrenaline and other agents which stimulate adrenergic receptors also produce effects similar to those described above for periarterial nerve stimulation. Therefore, this preparation can be used to study nerve-mediated as well as direct muscle-stimulated adrenergic responses.
4. It is a convenient method for differentiating between drugs with actions at the adrenergice neurone (adrenergic neurone blockers), and those acting post-junctionally at adrenergic receptors. Agents that enhance the effect of periarterial nerve stimulation or the action of noradrenaline such as tyramine, amphetamine or uptake$_1$ inhibitors, e.g. cocaine (from

Erythroxylon coca) and imipramine, are referred to as *sympathomimetics*. Those agents which inhibit periarterial nerve stimulation and the effects of noradrenaline, such as α adrenergic receptor antagonists, e.g. ergocristine (from *Claviceps purpurea*), prazosin, phentolamine or propranolol, are called *sympatholytics*. Classical pharmacological theory regards the adrenergic receptors mediating jejunum relaxation as a mixture of α- and β-adrenoceptors. Some recent studies suggest that they might be adrenergic β-adrenoceptors. Agents which act prejunctionally to inhibit the contractions of the IRJ by preventing the release of neurotransmitter, such as guanethidine, are called *adrenergic neurone blockers*.

SETTING UP THE ISOLATED RABBIT JEJUNUM (FINKLEMAN PREPARATION) (FIGURE 10.3)

The same considerations described for the GPI apply. The IRJ is best worked at 37 °C, and the PSS should be aerated with a gas mixture of 95% oxygen and 5% CO_2 to maintain a pH of about 7.4. The IRJ preparation is set up essentially as first described by Finkleman (1930) and employed by Boura and Green (1965) to discover the mode of action of adrenergic neurone blockers. Again, the length of muscle must be carefully determined in order to ensure that reproducible results are obtained.

Procedure

A length of IRJ is obtained with the mesenteric vessels attached. The sympathetic nerves supplying the muscle run along the blood vessels. The attached mesentery is carefully trimmed and threaded through an electrode (Burn and Rand type), which enables the periarterial nerves to be stimulated. Rectangular pulses of 25–40 V, 0.5 ms pulse width, at a frequency of 5–25 Hz delivered for 5–10 s should produce frequency-dependent relaxations, similar to dose-dependent responses to noradrenaline or adrenaline:

1. The plant extract may relax the IRJ in a manner similar to noradrenaline (NA). If such a response, as that to NA, is blocked by phentolamine, propranolol or a mixture of these adrenergic receptor antagonists, an effect at the adrenoceptor is suggested. If the response is resistant to these antagonists, the extract may be acting directly on the contractile mechanism, such as activation of second messengers, block of Ca^{2+} channels or opening of K^+ channels.
2. An adrenergic neurone blocking property is likely if the extract does not relax the IRJ directly, but blocks the effect of periarterial nerve stimulation, with no effect on the response to NA or adrenaline added to the bath.
3. The extract, without directly relaxing the muscle, may potentiate the response to nerve stimulation and NA. In that case, an uptake inhibitor action like that of cocaine is indicated.

Adrenergic neurone blockers are sometimes used in the treatment of hypertension; bretylium is used as an anti-arrhythmic.

Skeletal Muscle–Motor Nerve Transmission

Impulses from the motor nerve to skeletal (voluntary) muscle are transmitted by ACH. Although the nerve lies in close proximity to the muscle which it innervates, there is no fusion. The gap between the nerve ending and the motor end-plate is called the neuromuscular junction (NMJ). This synaptic gap exists in all nerve–muscle junctions, but NMJ is more conventionally used to describe the skeletal muscle–motor nerve junction. Drugs which interfere with NMJ transmission are used as muscle relaxants; some such drugs act centrally, e.g. mephenesin, diazepam; the method described here is for drugs with peripheral activity at the motor end-plate and are of two main types:

1. Persistent motor end-plate depolarisers, e.g. suxamethonium.
2. Blockers of ACH receptors at the motor end-plate on the lock and key principle, e.g. tubocurarine (from *Chondodendron tomentosum*), and toxiferine (from *Strychnos toxifera*).

The other group of drugs acting at the NMJ with clinically useful applications are anticholinesterases, e.g. physostigmine (from *Physostigma venenosum*) and several synthetic derivatives, such as neostigmine. These are used in the treatment of myasthenia gravis.

THE RAT PHRENIC NERVE–DIAPHRAGM PREPARATION (FIGURE 10.5)

This is a well known preparation for the study of the action of muscle relaxants, first described by Bulbring (1946). An adult rat is killed by a stunning blow on the head and bled. The skin is removed over the chest and the thorax opened along the right side of the sternum. The frontal part of the right thoracic wall is removed, when the phrenic nerve can be seen. Both left lobes of the lungs are carefully removed. The end of the nerve distal to the diaphragm is freed from connective tissue, cut and tied securely with cotton thread. The rest of the diaphragm is freed, taking care not to damage the nerve. The isolated preparation consisting of nerve and one half of the diaphragm is trimmed to a triangle.

The whole preparation is mounted on a special phrenic nerve electrode, in which the broad end of the triangle is tied at two points with the diaphragm muscle lying along the length of one end of the platinum electrode. The nerve is looped through the other end of the electrode. The apex (tendinous end) of the triangle is attached to a recording device. In this way the muscle can be stimulated directly and indirectly through the nerve. Care should be taken that the nerve is left slack. The preparation should be mounted in PSS, aerated with a gas mixture of 95% O_2 and 5% CO_2. The nerve and muscle are stimulated with rectangular pulses, supramaximal voltage (20–40 V, pulse widths of 0.5–2.0 ms at a frequency of 1 Hz.

A plant extract may cause:

1. Inhibition of twitch responses to both direct and indirect stimulation of diaphragm muscle. This would be a non-specific effect. Such an effect is of little therapeutic potential but may explain toxic effects of plant extracts.
2. Inhibition of indirect but *not* direct stimulation, that is, an action at the NMJ. This is of potential therapeutic value, as all the muscle relaxants in therapeutic use act at this site. If the block is reversed by anticholinesterases (e.g. physostigmine or neostigmine), then a competitive type of block similar to tubocurarine is indicated. If not, a depolarising type of block such as by suxamethonium is indicated.
3. Reversal of the block caused by tubocurarine, but *not* that by suxamethonium; an anticholinesterase type of action is indicated.

The Frog (or Toad) Rectus Abdominis Muscle Preparation

The rectus abdominis of the frog is skeletal muscle consisting of two types of muscle fibre—fast twitch-producing fibres and slow contracture-producing fibres. Both types of fibres are also present in some mammalian skeletal muscles. Acetylcholine causes a *contracture* (slow developing, but sustained contraction) of the rectus abdominis muscle. The rectus abdominis muscle preparation can be used to study the action of drugs on skeletal muscle. Drugs such as suxamethonium which paralyse mammalian skeletal muscle by depolarisation, cause a contracture by stimulating ACH receptors on the slow muscle fibres. In contrast, drugs such as tubocurarine, which paralyse mammalian skeletal muscle by competitive antagonism with ACH receptors at the motor end-plate, do not contract the rectus muscle, but antagonise the contracture caused by ACH or suxamethonium.

Procedure

The rectus preparation is easy to make. The frog is killed with a stunning blow to the head, followed immediately by pithing. The animal is placed on its back. When the skin is removed, the rectus abdominis is found lying along the length of the middle of the abdomen in two bundles. One half is set up in an organ bath containing a PSS such as frog Ringer in the usual way. The muscle is robust, and a resting tension of up to 2 g may be applied. In the tropics, the experiment can be done at room temperature. A plant extract may:

1. Produce a contracture. If the response is blocked by tubocurarine, a depolarising type of muscle relaxant action is indicated.
2. Produce no contracture, but may potentiate the response to ACH. An anticholinesterase type of action is indicated.
3. Produce no contracture, but inhibit the response to ACH. A tubocurarine type of action is indicated.

Experimental Set-up and Treatment of Results

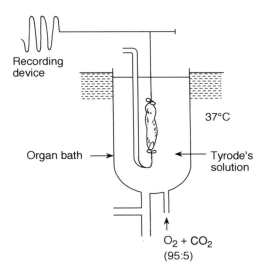

Figure 10.2   Guinea-pig ileum (GPI)

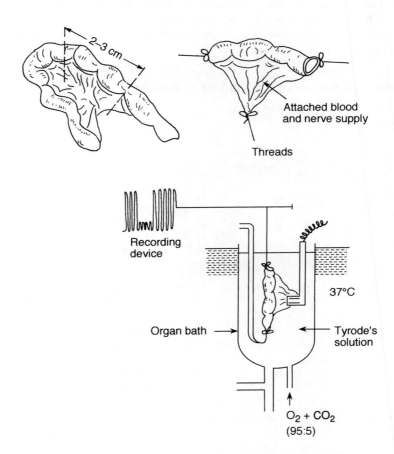

Figure 10.3   The isolated rabbit jejenum (IRJ) or Finkelman preparation.

Isolated Rabbit Jejunum: Inhibition of Spontaneous Movement by Plant Extract and Verapamil—Examples of Traces Obtained

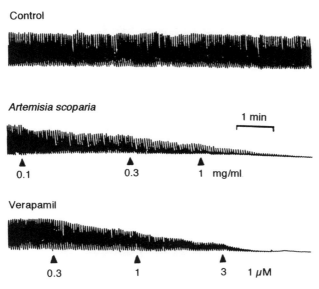

Figure 10.4   A representative tracing showing inhibitory effects of *Artemisia scoparia* and verapamil on spontaneous movements of isolated rabbit jejunum. Triangles represent the time at which the drugs were added to the tissue bath (10 ml) and normal saline (0.1 ml) was administered at respective times in upper panel (control). From Ca^{++} channel blocking activity of *Artemesia scoparia* extract. Gilani AH *et al* (1994), *Phytother. Res.*, **8**(3), 161–165, with permission.

Rat Phraenic Nerve–Hemidiaphragm Preparation: Inhibition of Twitch Response to Muscle Stimulation by Plant Extract—Example of Traces Obtained

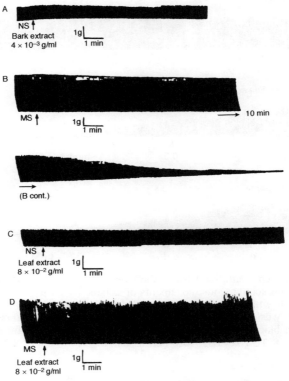

Figure 10.5 Typical recordings of the effect of *Bridelia ferruginea* extract on indirect (via phrenic nerve, NS) and direct (on the muscle, MS) electrically evoked twitch contractions of rat phrenic nerve–hemidiaphragm muscle preparation. Activity profile of bark extract (4×10^{-3} g/ml) added at the arrow on NS (A) and MS (B) are shown. Note the slow onset of MS depression by the bark extract. Activity profile of leaf extract (8×10^{-2} g/ml) added at the arrow on NS (C) and MS (D). From Effects of stembark and leaf extracts of *Bridelia ferruginea* on skeletal muscle. Onoruvwe O *et al* (1994), *Phytother. Res.*, **8**(1), 38–41, with permission.

THE EFFECT OF PLANT EXTRACTS ON THE CENTRAL NERVOUS SYSTEM

Introduction

It is often necessary to test a plant extract for its action on the central nervous system (CNS), to confirm its reputation in traditional usage or simply to complete the pharmacological profile of the extract. CNS action may be the main focus of interest for a particular laboratory in studying plant extracts, bearing in mind that many CNS-acting drugs currently used in medicine, or abused, are extracted from plants. Because of continuing advances in neuropharmacology, diseases of CNS origin, such as disorders of movement, neuroses and psychoses, can now be traced to biochemical lesions in specific areas of the brain, and to specific abnormalities in transmitter function, synthesis or release. Such investigations are beyond the scope of this book. What can be said here is that CNS-acting drugs can be broadly classified as in Table 10.2, which gives examples intentionally slanted to drugs of plant origin.

This does not mean that all drugs used in the treatment of CNS diseases can be accommodated by this classification. For instance, anticholinergics and carbidopa, used in the treatment of Parkinsonism, do not fall into any of these classes. It can be seen from the examples given that CNS activity is present in many plant extracts. Methods for testing are summarised here and referenced, and analgesic activity discussed in Chapter 8.

Table 10.2 Classification of drugs of plant origin acting on the CNS

Class	Drug example	Source
Stimulants	Cocaine	*Erythroxylum coca*
	Khat (cathine, cathinone)	*Catha edulis*
	Cannabis (tetrahydrocannabinol)	*Cannabis sativa*
	Amphetamine	Synthetic
Depressants	Barbiturates (nembutal, tuinal)	Synthetic
	Benzodiazepines (diazepam)	Synthetic
	Opioids (morphine, heroin)	*Papaver somniferum*
Antidepressants	Imipramine	Synthetic
	Deprenyl	Synthetic
Tranquillisers	Benzodiazepines (anxiolytics)	Synthetic
	Chlorpromazine (neuroleptic)	Synthetic
	Reserpine	*Rauvolfia* spp.
Hallucinogens	Lysergic acid diethylamide (LSD)	Derivative of ergot (*Claviceps purpurea*)
	Mescaline	*Lophophora williamsii* (cactus)
	Magic mushroom (psilocin)	*Psilocybe mexicana* (fungus)

Some Commonly Used Rodent Methods in Screening for CNS Activity (Figures 10.6–10.7; Tables 10.3–10.5)

Locomotor Activity of Mice

This test involves placing a number of mice in an *activity cage* which enables movement of the animals across a light beam to be recorded as a locomotion count (*see* Svensson and Thieme, 1969). Activity cages are commercially available. This test can be used to measure the effect of an extract on spontaneous activity or on caffeine or amphetamine-induced hypermotility (Bushnell, 1986). This test will demonstrate a CNS depressant or stimulant activity profile. A reduction in the CNS stimulant activity of *d*-amphetamine is indicative of antipsychotic potential.

Locomotor Co-ordination (Revolving Bar Test)

Mice are individually placed on a rotating bar (1 cm diameter; 6 turns/min; time of a single exposure, 2 min). Animals are preselected so that only those staying on the rod for this period are used. For further details *see* Boissier *et al* (1972).

Barbiturate Sleeping Time

Barbiturates (pentobarbitone, 55 mg/kg, or hexobarbitone, 35 mg/kg) are used to induce sleep. The time between loss and recovery of *righting reflex* is recorded as sleeping time. A stimulant or depressant effect of an extract on the CNS is indicated by a reduction or prolongation of the barbiturate-induced sleeping time.

Pentetrazole-Induced Convulsions in Mice

Pentetrazole (80–120 mg/kg) given intraperitoneally, causes clonic convulsions in mice. The following parameters are evaluated for anticonvulsant activity of a plant extract: time of onset of convulsion after drug administration; number of animals convulsing; number of convulsive episodes during a 10 min observation period; and number of animals dying in the group (for further details *see* Swinyard *et al*, 1952).

Other Tests

Newer tests for antidepressant activity will be covered in Volume 2.

Treatment of Results

Locomotor Activity of Mice Stimulated with Amphetamine or Caffeine

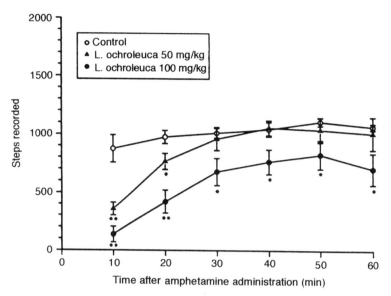

Figure 10.6   Effect of *Laminaria ochroleuca* extract on amphetamine-induced hypermotility. Each point represents the mean ±SEM of 5–6 groups of animals (3 mice/ group). Significance of differences with respect to the control group was evaluated by the Mann–Whitney U-test (*$p < 0.05$; ** $p < 0.01$). From *Laminaria ochroleuca*: a preliminary study of its effect on the CNS. Vasquez-Freire MJ *et al* (1994), *Phytother. Res.*, 8(7), 422–425, with permission.

Table 10.3   Influence of the BN fraction on locomotor activity of mice stimulated with caffeine. Time of observation 30 min after BN administration, $n = 8$

| Compound | Dose (mg/kg) | Average number of movements | | |
		$\times$ + SEM	%	ED$_{50}$ (mg/kg)
–	–	409.125 + 35.95	100	
BN	100	297.25 + 26.5[a]	72.65	167.5
	200	178.125 + 13.04[b]	43.54	108.1–259.6)
	300	97.5 + 15.27[b]	23.83	

[a] $p < 0.01$, [b] $p < 0.001$. BN = an active fraction. From Pharmacological properties of a lyophilizate from *Galeopsis tadanum* on the CNS of rodents. Czarnecki; R *et al* (1993), *Phytother. Res.*, 7(1), 9–12, with permission.

Locomotor Co-ordination: Revolving Bar Test

Table 10.4   Effect of *Cystoseira usneoides* extract on motor co-ordination

Extract dosage	30 min p.a.		60 min p.a.	
(mg/kg)	Time-on-rod (sec)	Failures	Time-on-rod (sec)	Failures
0 (control)	180.0 ± 0.0	0.0 ± 0.0	180.0 ± 0.0	0.0 ± 0.0
6.25	155.4 ± 24.6	0.8 ± 0.8	161.8 ± 15.0	0.8 ± 0.5
12.5	108.6 ± 33.0^a	2.3 ± 0.5^b	142.0 ± 16.3	1.5 ± 0.6^a
25	70.5 ± 17.2^b	3.2 ± 0.5^b	52.0 ± 14.7^b	3.4 ± 0.2^b

Each value represents the mean $\pm$ SEM of 4 groups (4 mice/group). Significance of differences with respect to the control group was evaluated by the Mann–Whitney U test (time-on-rod) or the Chi-square test (failures) ($^a p < 0.05$; $^b p < 0.01$). From Neuropharmacological effects of *Cystoseira usneoides* extract. Vazquez-Freire MJ et al (1995), *Phytother. Res.*, 9(3), 207–210, with permission.

Pentobarbital-induced Hypnosis (Barbiturate Sleeping Time)

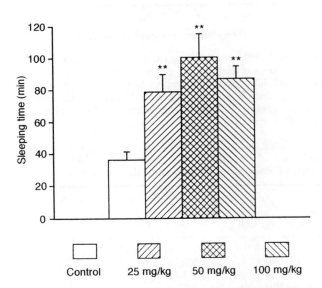

Figure 10.7   Effect of *Laminaria ochroleuca* extract on pentobarbital-induced hypnosis. Each bar represents the mean $\pm$ SEM of 7–8 animals. Significance of differences with respect to the control group was evaluated by the Mann–Whitney U-test (** $p <$ 0.01). From *Laminaria ochroleuca*: a preliminary study of its effect on the central nervous system. Vazquez-Freire MJ et al (1994), *Phytother. Res.*, 8(7), 422–425, with permission.

Pentetrazole-induced Convulsions in Mice

Table 10.5 Influence of the BN fraction on convulsions induced with pentetrazole in mice. Time of observation 30 min after BN administration, $n = 6$

Compound	Dose (mg/kg)	Time convulsions appeared [sec]		% of animals in which convulsions appeared	Number of attacks during 10 min	
		$\times$ + SEM	%		$\times$ + SEM	%
–	–	79 + 8.86	100	100	2.67 + 0.49	100
BN	100	113 + 15.15	142.1[a]	100	2.5 + 0.56	93.6
	200	132 + 14.11	166.1[b]	83.3	0.83 + 0.17	31.1[c]

[a] $p < 0.05$; [b] $p < 0.02$; [c] $p < 0.01$, BN = plant fraction. From Pharmacological properties of a lyophilizate from *Galeopsis ladanum* on the central nervous system of rodents. Czarnecki R *et al* (1993), *Phytother. Res.*, 7(1), 9–12, with permission.

FURTHER READING

Peripheral Nervous System

Methods

Arunlakshana O and Schild HO (1959). Some quantitative uses of drug antagonists. *Br. J. Pharmacol.*, **14**, 48–58.

Boura ALA and Green AF (1965). Adrenergic neurone blocking agents. *Ann. Rev. Pharmacol.*, **5**, 183–212.

Brown J *et al* (1980). Effects of α-adrenoceptor agonists and antagonists and of antidepressants on pre- and post-synaptic α-adrenoceptors. *Br. J. Pharmacol.*, **67**, 33–40.

Bulbring E (1946). Observations on the isolated phrenic nerve diaphragm of the rat. *Br. J. Pharmacol.*, **1**, 38–61.

Finkleman B (1930). On the true nature of inhibition of the intestine. *J. Physiol. Lond.*, **70**, 145–157.

Kimura M *et al* (1987). Pharmacological evidence for an interaction between constituents of the Japanese-Sino medicine 'Keishi-ka-zyutubu-to' in neuromuscular blockade in diabetic mice. *Phytother. Res.*, 1(3), 107–113.

Long JP and Chiou CY (1970). Pharmacological testing methods for drugs acting on the peripheral nervous system. *J. Pharm. Sci.*, **59**, 144.

Paton WDM and Zar MA (1968). The origin of acetylcholine released from guinea-pig intestine and longitudinal muscle strips. *J. Physiol. Lond.*, **194**, 13–33.

Sen T and Chaudhuri AKN (1992). Studies on the neuropharmacological aspects of *Pluchea indica* root extract. *Phytother. Res.*, 6(4), 175–179.

Related Pharmacological Experiments

Benedicta NN *et al* (1993). Anticholinergic effects of the methanol stembark extract of *Erythrina sigmoidea* on isolated rat ileal preparations. *Phytother. Res.*, 7(2), 120–123.

Birmingham AT *et al* (1970). Comparison of the sensitivities of innervated and denervated rat vasa deferentia to agonist drugs. *Br. J. Pharmacol.*, **30** 748–754.

Blakeley AG *et al* (1988). A study of the actions of P_1-purinoceptor agonists and antagonists in the mouse vas deferens *in vitro*. *Br. J. Pharmacol.*, **94**, 37–46.

Burnstock G (1972). Purinergic nerves. *Pharmacol. Rev.*, **24**, 509–581.

Burnstock G *et al* (1972). Atropine-resistant excitation of the urinary bladder: the possibility of transmission via nerves releasing a purine nucleotide. *J. Pharm. Pharmacol.*, **44**, 668–688.

Dumsday B (1971). Atropine resistance of the urinary bladder. *J. Pharm. Pharmacol.*, **23**, 222–225.

Dunham NW and Miya TS (1957). A note on a simple apparatus for detecting neurological deficit in rats and mice. *J. Am. Pharm. Sci. Ed.*, **46**, 208–209.

Pagala MKD (1983). An *in vitro* fluid electrode technique to record compound muscle action potential along with nerve-evoked tension. *J. Electrophysiol. Tech.*, **10**, 111–118.

Pagala MKD (1987). An *in vitro* electromyography chamber to monitor neuromuscular function. *J. Electrophysiol. Tech.*, **14**, 197–209.

Prabhakar E and Nanda Kumar NV (1992). Potentiating action of *Datura metel* root extract on rat intestinal cholinesterase. *Phytother. Res.*, **6**(4), 160–162.

Udoh F (1995). Effects of leaf and root extracts of *Nauclea latifolia* on purinergic neurotransmission in the rat bladder. *Phytother. Res.*, **9**(4), 239–243.

Central Nervous System

Methods

Boissier JR and Simon P (1960). Planche à trous automatisee. *Therapie*, **15**, 1170–1174.

Boissier JR *et al* (1972). Étude psychopharmacologique experimentale d'une nouvelle substance psycotrope, la 2-ethylamino-6-chloro-4-phenyl-4,3,1-benzoxacine. *Therapie*, **72**, 325–338.

Bushnell PJ (1986). Differential effects of amphetamine and related compounds on locomotor activity and metabolic rate in mice. *Pharmacol. Biochem. Behav.*, **25**, 161–170.

Dandyia PC and Callumbine H (1959). Studies on *Acorus calamus*: pharmacological actions of essential oils. *J. Pharmac. Exp. Ther.*, **125**, 353–359.

Glowinsky J and Iverssen LL (1966). Regional studies of catecholamines in rat brain. 1. The disposition of [^{3}H] norepinephrine, [^{3}H] dopamine and [^{3}H] dopa in various regions of the brain. *J. Neurochem.*, **13**, 656–669.

Goodman LS *et al* (1953). Comparison of maximal seizures evoked by pentylenetetrazol (Metrazol) and electroshock in mice and their modification by anticonvulsants. *J. Pharmacol. Exp. Ther.*, **108**, 168–176.

Knoll J *et al* (1961). Beta-aminoketones, a new group of tranquillizers. *Arch. Inst. Pharmacodyn. Ther.*, **130**, 155–169.

Laguna MR *et al* (1993). Effects of extracts of *Tetraselmis suecica* and *Isochrysis galbana* on the central nervous system. *Planta Medica*, **59**, 207–214.

Svensson TH and Thieme G (1969). An investigation of a new instrument to measure motor activity of small animals. *Psychopharmacology* (Berl.), **14**, 157–163.

Swinyard EA *et al* (1952). Comparative assays of antiepileptic drugs in mice and rats. *J. Pharmacol. Exp. Ther.*, **106**, 319–330.

Vazquez-Freire MJ *et al* (1995). Neuropharmacological effects of *Cystoseira usneodes* extract. *Phytother. Res.*, **9**(3), 207–210.

Related Experiments and Methods

Boschi G *et al* (1987). Neuroleptic-induced hypothermia in mice: a lack of evidence for a central mechanism. *Br. J. Pharmacol.*, **90**, 745–751.

Janssen PAJ *et al* (1960). Chemistry and pharmacology of compounds related to 4-(4-hydroxy-4-phenyl-piperidino)-butyro-phenone. Part IV. Influence of haloperidol and of chlorpromazine on the behaviour of rats in an unfamiliar 'Open Field' situation. *Psychopharmacologia*, **1**, 389–392.

Karoum F and Egan MF (1992). Dopamine release and metabolism in the rat frontal cortex, nucleus accumbens, and striatum: a comparison of acute clozapine and haloperidol. *Br. J. Pharmacol.*, **105**, 703–706.

Magnusson O *et al* (1987). Effects of dopamine D_2 selective receptor antagonist remoxipride on dopamine turnover in the rat brain after acute and repeated administration. *Pharmacol. Toxicol.*, **60**, 368–373.

11

Endocrine Activity: Antifertility and Sex Hormones

Plant drugs have been used since time immemorial for their effects upon sex hormones, in particular for suppressing fertility, regularising the menstrual cycle, relieving dysmenorrhoea, treating the enlarged prostate, menopausal symptoms, breast pain and during and after childbirth. Although there is a lot of anecdotal evidence, scientific proof for many of these uses is hard to find. Examples with supporting evidence include: gossypol (from *Gossypium hirsutum*) as an antifertility agent in the male; embelin (from *Embelia ribes*) which is abortifacient in the female; agnus castus (from *Vitex agnus-castus*) and evening primrose oil (from *Oenothera biennis*) for premenstrual syndrome and breast pain. Saw palmetto berries (*Sabal serrulata*) are taken for prostate enlargement, and of course ergometrine from ergot (*Claviceps purpurea*) is the standard oxytocic drug for use after childbirth.

Apart from plant medicines being taken for their known properties, there are numerous examples of plants in the diet having accidental effects. This is how gossypol was discovered as a contraceptive, when cooking with cottonseed oil led to greatly reduced male fertility in parts of China. More recently, the lower rate of enlarged prostate and prostate cancer in Japanese men has been linked to the ingestion of plant oestrogens in soya. These phyto-oestrogens are found widely in the Leguminosae, and are also responsible for affecting milk levels and reducing fertility in animals ingesting them in forages. The condition 'Clover disease', in which infertility in sheep is accompanied by cystic glandular hyperplasia of the cervix and uterus, is due to ingestion of *Trifolium pratense* (clover) which contains oestrogenic isoflavones. These are also found in alfalfa (lucerne), *Medicago sativa*.

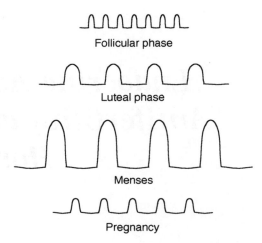

Figure 11.1   Response of the uterus to oxytocic drugs at different phases of the oestrus cycle.

Physiology of the Uterus

The physiology of the uterus and its response to oxytocic drugs differs greatly in different species. The type of motility and the threshold for the response to oxytocic drugs differs with the phase of the oestrus cycle and the stage of pregnancy. During the *follicular* or proliferative phase, contractions are short in duration and develop only low pressures. During the *luteal* secretory phase, contractions are less frequent and are longer in duration, lasting 1–2 min. With menses, the strength and to some extent the frequency of the contractions increase. In the presence of pregnancy the tone and contractility of the uterus are markedly reduced initially. However, during the fourth or fifth month of gestation, contractility increases slightly, although it is only as time of delivery approaches that regular strong contractions appear (*see* Figure 11.1).

These variations in contractility are due to the complex regulatory mechanisms to which the uterus is exposed. These include endocrine, neural, mechanical and chemical regulations.

The sensitivity of the uterus can be increased by pretreating the animal with an oestrogen, usually stilboestrol with a dose of about 1 mg/ml 24 h before the experiment. However, this may not be necessary for bioassay studies, but the uterus status should be determined.

Spontaneous Activity of the Uterus

The innervation of the uterus is very scanty and there are probably no ganglion cells as in other smooth muscles. It has been suggested that hormonal control of the uterus is more important than neural control in its regulation. It is also reported that the number of receptors for drugs are under hormonal control. The uterus, both *in situ* and when excised, contracts rhythmically. These contractions are myogenic in origin since they are not abolished by section of the uterine nerves.

The fundus is the origin of these spontaneous contractions and myometrial cells in this area function as pacemakers, giving rise to conducted action potentials. During pregnancy, electrical and mechanical quiescence and relative inexcitability are produced in the uterus by progesterone acting locally at the site of implantation of the foetus. The frequency and force of contractions increases gradually as pregnancy advances and become fully co-ordinated during parturition, at which time progesterone is withdrawn and labour sets in.

Contraceptive and Abortifacient Activity

The term *contraceptive* refers to a substance or device used in the prevention of pregnancy, by any mechanism. If, however, the pregnancy is established and then terminated, the substance is termed *abortifacient*.

The term *postcoital contraception* (or interception) applies to situations where mating has taken place but fertilisation may or may not have occurred. It therefore describes agents which prevent implantation of fertilised ova. This could be considered to be a form of early abortion, and must be borne in mind, since many people who accept the need for contraception cannot accept abortion on principle.

Contraceptives include hormonal agents which prevent ovulation or interfere with fertilisation by altering the vaginal or uterine environment; cytotoxic agents which affect development of ova and spermatozoa; and mechanical and chemical barriers to fertilisation. Post-coital contraceptives at present are hormonal or mechanical (e.g. emergency insertion of an intra-uterine device). Apart from some barrier methods, most forms of contraception are directed at the female, although the search for male contraceptives continues. (The main problem is that the hormones in men responsible for development of sperm are also necessary for maintenance of libido, and loss of libido is not usually considered an acceptable method of contraception.)

Abortifacients work by several methods. These include stimulation of the uterine muscle and therefore expulsion of the contents; antagonism to progesterone which would maintain the pregnancy; and cytotoxicity, which kills the developing foetus. Cytotoxicity is rarely acceptable on safety grounds, since if abortion does not occur teratogenicity may result.

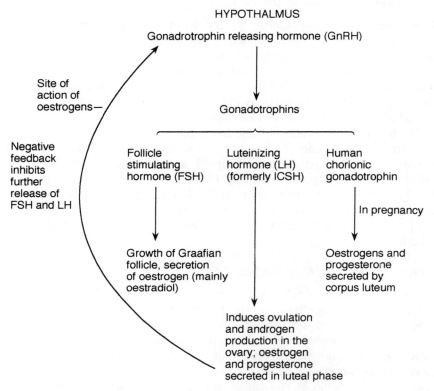

Figure 11.2   Sex hormones: relationships and effects.

Effects of oestrogens in the female:

- Stimulation of secondary sex characteristics at puberty
- Prevention of ovulation by negative feedback (use in contraception)
- Stimulation of endometrial proliferation
- Decrease bone reabsorption (use in HRT for osteoporosis)
- Sensitises uterus to contractile agents, e.g. prostaglandins.

TESTING FOR HORMONAL AND ANTIFERTILITY EFFECTS

There are conditions where pharmacological evaluation of activity is difficult if not impossible, where it is not applicable to animals, and clinical studies have to be used. These include premenstrual syndrome, cyclical mastalgia, menopausal disorders, facilitating childbirth and so on. In some disorders non-hormonal treatment may be preferred, for example dysmenorrhoea (painful menstruation) is usually treated with cyclo-oxygenase inhibitors, since it is caused by prostaglandin stimulation of the oestrogen-sensitised uterus.

Antifertility estimation in the female measures the pregnancy rate and includes anti-ovulation, anti-implantation and cytotoxic agents. In the male it includes inhibition of spermatogenesis or sperm motility, and again cyto-toxicity, assuming of course that libido is not affected and mating still takes place. Reproductive physiology is too complex to deal with here, but an outline of the natural production and effects of oestrogens is shown in Figure 11.2. It would be wise to test all those plants found to have antifertility activity for oestrogenicity. Compounds such as finasteride, which inhibit the enzyme 5-α-reductase, are now being used to treat benign hyperplasia of the prostate and methods are referenced.

It is also possible to measure oxytocic activity, using isolated rat uterus or uterine strips, which would indicate a use during and after childbirth or as an abortifacient.

The following experiments will therefore be described: a protocol for anti-fertility testing; determination of oestrus and mating; antifertility and anti-implantation methods; abortifacients; and oxytocic and oestrogenic activity.

Animal Models: Antifertility Testing

A basic protocol for the testing of plant extracts for effects on female rats is given; this can be adapted to other species as long as differences in reproduc-tive cycles are taken into account (for full review, *see* Farnsworth *et al* 1975).

Methods for testing anti-implantation and abortifacient activity are very similar, but rats in the second or third trimester of pregnancy are used for the latter. Mechanisms for both may be different; substances preventing implantation may have no effect on an established pregnancy.

Uterine stimulation (oxytocic action) may be investigated using strips of uterine smooth muscle in an organ bath; dose-dependent contractions are obtained from a substance showing oxytocic activity. Usually experiments are performed using both pregnant and non-pregnant uterine preparations because they have differing sensitivities.

Male contraceptive activity testing involves measuring sperm count and motility in treated animals, and retesting after withdrawal of treatment to ensure reversibility. This is referenced only.

Oestrogen-containing contraceptives which prevent ovulation are the basis of the combined contraceptive pill. The post-coital or 'morning after' pill contains oestrogen at a higher dose and works by preventing implantation. Because of this wide usage, and because dietary ingestion of oestrogenic sub-stances is so important, oestrogenic activity will be dealt with separately.

As progesterone is essential for maintaining pregnancy, any agent which blocks its effect (such as mifepristone) will induce abortion. Administration of plant extract followed by estimation of plasma progesterone levels will indicate this type of effect.

Testing for Antifertility Activity in Female Rats: Protocol Example*

Animals must be housed singly (for obvious reasons) and acclimatised for 1–2 weeks. Check vaginal smears after 10–14 days to indicate normal oestrus cycle (length of cycle in unmated rats is 4–5 days).

1.   Dose virgin female rats daily with vehicle or plant extract and examine for changes, to indicate:
   - *Oestrogenicity*, e.g. cornification of vagina, increase in weight of uterus (*see* method).
   - *Anti-oestrogenicity*, which can be tested by administration with standard oestrogen to see if effects blocked (*see* 'Method').
   - *Anti-gonadotrophic activity*, tested by measuring serum and pituitary hormone levels.

2.   Pair animals with males which are proven breeders and continue daily dosing; examine vaginal smears for evidence of mating (*see* Method).

3(a).   If no mating: possible lack of oestrogen and/or progesterone due to peripheral antagonism or central inhibition (antigonadotrophic activity). For further procedure, *see* 6 below).

3(b).   If mating occurs, continue dosing one group of rats (no smears) for about 7 days, then examine oviducts, ovaries and uteri for:
   - *Ovulation*. If this does not occur, may indicate CONTRACEPTIVE activity.
   - *Fertilisation*. If this occurs, examine ova to see whether transportation and implantation occur. Look for luteolytic effects. If any of these are altered, may indicate ANTI-IMPLANTATION (INTERCEPTIVE) activity, also known as post-coital contraception. Check for cytotoxic effects on the ova, including teratogenicity.

4.   After mating, continue dosing second group of rats up to and including day 19. Check for:
   - *Lack of weight gain*: indicates CONTRACEPTION or ANTI-IMPLANTATION
   - *Bleeding, loss of weight gained*, due to abortion or resorption: indicates ABORTIFACIENT activity. If abortifacient, check for *oxytocic* activity (uterine stimulation) (*see* 'Method')

---

*Adapted from Potential value of plants as sources of anti-fertility agents I and II. Farnsworth N *et al* (1975). *J. Pharm. Sci.*, **64**, 535–596 and 717–754, with permission.

5. Sacrifice and post-mortem this group on day 20. Look for and compare with controls: *Number of normal living foetuses, corpora lutea of pregnancy, number of abnormal foetuses* (living or not), together with resorbing foetuses and implantation sites.

6. Non-mating group: cease dosing at day 20 and continue to house with males for a further 20 day period. If still no mating, some permanent effect on fertility may have occurred. Sacrifice and perform post-mortem, examining reproductive organs and pituitary gland histologically. Measure serum and pituitary hormone levels. If mating does occur, repeat from 3, 'Fertilisation'.

Materials and Methods (Figures 11.3–11.5; Tables 11.1–11.3)

Determination of Oestrus and Ovulation; Mating and Pregnancy

Preliminary experiments using virgin rats treated with plant extract may be carried out to see if ovulation is inhibited, suggesting contraceptive action. Vaginal smears are taken each morning and examined under a light microscope; the presence of nucleated epithelial cells indicates the advent of ovulation. For subsequent fertility testing, rats of proven fertility may be used. If the length and frequency of the oestrus cycle is under investigation, changes in these cells can be monitored using the methods cited. The rat is a spontaneous ovulator with a regular oestrus cycle of 4–5 days (in contrast to the rabbit, which is a coitus-induced ovulator). Mating is allowed at the proestrus stage or when ovulation has been demonstrated; males of proven fertility are allowed into the cage overnight at a ratio of between 4:1 and 2:1 female to male. The presence of groups of spermatozoa in the vaginal smear the next morning indicates that mating has taken place, and this is designated Day 1 of pregnancy. The pregnant rats are then divided into two or more groups (of at least five per group) for treatment with plant extract and as a control.

Antifertility and Anti-implantation in the Female Rat

Rats are dosed orally each day with plant extract, and the control group with vehicle, through days 1–5 of pregnancy. Laparotomy is then carried out on day 10 or thereabouts, under light ether anaesthesia, to record the number of implants. These animals may then be allowed to carry the pregnancy to term, and the litter observed for numbers, deformities, etc. Alternatively the animals may be sacrificed, making examination of implantation and resorption sites easier.

The number of females showing no implantation is also recorded and an index for antifertility or anti-implantation may be expressed in the following way:

$$\text{Antifertility activity} = \frac{\text{No. of animals showing no implantation}}{\text{Total number of animals}} \times 100$$

$$\text{Anti-implantation activity} = \frac{\text{No. of implants in control} - \text{No. of implants in test group}}{\text{No. of implants in control group}} \times 100$$

Variations in protocol may be used, e.g. the test extract given for varying periods to give an indication of when the antifertility effect is arising:

Control group   Vehicle alone days 1–9
Test group   (a) Extract days 1–3
                   (b) Extract days 4–6
                   (c) Extract days 7–9
                   (d) Extract days 1–7

Abortifacient Activity

Methods are similar to those for measuring rates of implantation or fertility, and similar quantification may be used, but the pregnant rat is treated with plant extract during the second or third trimester, i.e. after 9 days. Estimation is carried out by examining the animal for vaginal bleeding on days 12–16, and by measuring increase in weight of the pregnant animal (or lack of it, if abortion or resorption of foetuses is occurring). Animals are sacrificed by cervical dislocation at about day 20 and the number of living or dead foetuses recorded.

Teratogenicity

Surviving foetuses from the above experiments are examined for deformities. If teratogenicity is indicated, experiments on this plant should be abandoned—unless a lower dosage level is to be investigated, in which case extra precautions to female personnel must be taken.

Measurement of Antigonadotrophin and Progesterone Levels

These are best carried out using commercially available radio-immunoassay (RIA) kits (e.g. from Organon), which will give direct measurement of hormone plasma levels. Treatment with plant extract may over a period of time affect these levels; in the case of progesterone a pregnancy will not be maintained if levels drop, and in the case of antigonadotrophic hormones the results are important from a mechanistic point of view, but are outside the scope of this book.

If an RIA kit for progesterone is not available, a bioassay method for measuring progestogenic and antiprogestogenic activity can be used, which utilises the traumatisation method of Ohta (1982) (*q.v.*) modified by Pathak and Prakash (1989).

Oxytocic Activity

In vitro

Oestrus status is determined (Short and Woodnatt, 1974). Pregnant uteri are obtained by mating animals in oestrus (rats or guinea-pigs) overnight with proven fertile males. Presence of spermatozoa in vaginal smears next morning indicates mating as before. Early pregnancy is taken as days 6–7; mid-pregnancy as 11–12 days. Uteri are obtained from animals which have been killed by a blow on the head and exanguinated. The abdomen is opened and the uterine horns severed at the junction with the fallopian tubes and placed in a dish of Tyrode or Jalon solution (*see* Appendix II). Pregnant uteri have strips of about 2 cm cut from the mid-portion of the horn, blastocysts having been gently removed. Non-pregnant uteri are cut into strips of about 1.5 cm. The strips of uterine tissue are suspended in an organ bath containing Tyrode solution and maintained at 36–37 °C and gassed with air, with an applied tension of about 0.5 g. Tissues should equilibrate for 30 min before application of drugs. Responses are monitored to obtain traces (Figures 11.3–11.4).

In vivo

Pregnant rats (determined as before) are divided into groups of at least four rats each, and groups treated as follows:

Control group	Vehicle only administered.
Test groups	Plant extract, dose level 1.
	Plant extract, dose level 2 etc.
Standard reference group	Oxytocin, 2 units/rat, s.c.

Method

Rats are weighed daily and results recorded from day 1. Drugs are given on days 15, 16 and 17 of pregnancy. After administration of drugs, animals should be observed for signs of abortion for 6 h. Fetal death *in utero* is considered to have occurred if weight remains constant or decreases. This may be confirmed by laparotomy on or about day 23 of pregnancy.

Oestrogenic Activity

Oestrogens, whether naturally produced by the body, obtained from the diet or taken as synthetic compounds, have a similar spectrum of activity. The term 'oestrogenic' describes these effects, rather than a chemical structure. Their action is complex and dose-dependent and is therefore discussed in general terms only. Sometimes effects appear to be conflicting, for example low doses of natural oestrogens in women protect the body from cardiovascular events such as stroke and myocardial infarction, but high doses may cause thrombosis; oestrogens stimulate the skeletal growth of girls at puberty (in conjunction with androgens) but later arrest the same process by closing the epiphyses.

In addition to these activities, oestrogens are responsible for the development and maintenance of secondary sex characteristics in women, stimulate endometrial proliferation, and decrease bone reabsorption. Some cancers are oestrogen-dependent (some breast and endometrial cancers). Oestrogens sensitise the uterus to prostaglandins, which may cause painful menstrual cramps, but synthetic oestrogens are used to treat that same condition by inhibiting ovulation. This is also the basis of the oestrogen-containing contraceptive pill, which affects the hormonal feedback mechanism. *See* Figure 11.2 for clarification; also 'Further Reading'.

In men, oestrogens suppress androgenic stimulation of the prostate and are used in the treatment of prostate cancer. The same principle would apply to enlargement of the prostate (benign hyperplasia of the prostate, BHP). Unfortunately, oestrogen-resistant cells may develop which limit the usefulness of this treatment. Oestrogen therapy in men may also lead to some unacceptable feminisation.

Plant oestrogens can have all of the above effects, whether taken as a medicine or by inadvertant ingestion, although the potency is less than the natural or synthetic hormones. At present the main interest in oestrogens is for their contraceptive effects, for hormone replacement therapy, and to prevent loss of bone density and other undesirable effects of oestrogen deficiency. In theory some of these weaker phyto-oestrogens could be utilised as anti-oestrogens, with the plant compound occupying receptor sites to the exclusion of the much more potent natural oestrogen. The three main groups are classified according to chemical structure: steroids, coumestans and isoflavonoids.

Steroidal phyto-oestrogens include oestrone and its tautomer, oestriol, which are the most potent. These are also natural human hormones. They are found in pomegranate, *Punica granatum*; liquorice, *Glycyrrhiza glabra*; kidney bean, *Phaseolus vulgaris*; and cereals such as oats, *Avena sativa*, wheat, *Triticum aestivum* and rice, *Oryza sativa*. Diosgenin, from *Dioscorea* and *Smilax*, and the ubiquitous β-sitosterol are also weakly oestrogenic.

Coumestrol is the major coumestan, found in *Glycine max* (soya), *Secale cereale*, *Taraxacum officinale* and *Trifolium* and *Medicago* species. Medicagol is found in *Medicago sativa*.

Isoflavonoids are found widely in the family Leguminosae, and include, in descending order of potency, genistein, daidzein, biochanin A and formononetin. These are abundant in *Genista, Baptisia, Trifolium, Cytisus* and *Medicago* spp. Prunetin is less common and occurs in *Prunus* species.

Other types of compounds are found in plants including anethole, found in fennel and anise (*Pimpinella anisum*), and asiaticoside, from *Centella asiatica*, which are weakly oestrogenic.

Animal Models: Testing for Oestrogenic Activity

Female rats or mice are usually used. When oestrogens are given to immature rats, a measurable increase in uterus size is obtained. If animals are ovariectomised, the resultant lack of oestrogen causes atrophy of the uterus; if an oestrogenic substance is then administered, the uterus will increase in size and after a period of time can be removed and weighed. This is usually quantified by selecting a suitable response, such as doubling in weight of the uterus, and compared with the dose of a standard (such as oestradiol) needed to produce the same response. This gives a measure of 'oestrogenic potency'.

Other changes can be noted at the same time, such as cornification of cells obtained from a vaginal smear, and the time at which they first appear related to a particular dose. This is the Allen–Doisey test.

Oestrogens also cause opening of the vagina, which can be used as a quantal response to a particular dose in a group of immature animals. All these parameters may be taken as indicative of oestrogenic activity.

A less frequently used model is that of delayed implantation, where mated animals (rats) are ovariectomised (on day 3 after mating) and the pregnancy maintained by giving progesterone. Implantation occurs under the influence of oestrogen, so the administration of a plant extract (usually on days 8 and 9) will lead to implantation if the test substance has oestrogenic activity. This is observed by laparotomy (on day 10). For other methods see Prakash *et al* (1992) and Prakash and Roy (1984) in the Further Reading list.

Anti-oestrogenicity can be assessed using any of the above models, by concurrent administration of test and standard oestrogen, to see if its effects are inhibited.

Materials and methods: Oestrogenicity and Anti-oestrogenicity (Figures 11.6–11.8; Table 11.4)

Immature rats or bilaterally ovariectomised animals are used.

Animals

1. Immature female rats: 21 days old, in groups of 6.

2. Bilaterally ovariectomised animals: Rats and mice (about 3 weeks old) are spayed under light ether anaesthesia, and used in groups of 5 or 6. Experiments are performed 7–15 days later for rats, 10 days for mice.

Test group: Plant extract plus vehicle used for standard reference.
Control group: vehicles only by same route.
Standard reference group: known oestrogen plus vehicle for extract.
Anti-oestrogenic test group: standard reference plus plant extract.

Note: The oestrogen is often given s.c., while the plant extract may be given p.o. or s.c. The vehicles for both must therefore be given in the control.

Reagents

Plant extract: 100 mg/kg p.o. rat; 20 mg/kg mouse; using a suitable range.
Standard oestrogens: Oestradiol benzoate 1 μg/10 g, in corn oil (0.1 ml) s.c.
Conjugated oestrogens 0.2 μg/kg mouse s.c.
Diethylstilboestrol (DES) 2 mg/kg in gum acacia p.o.

Method

Animals are administered substances under test for 3–5 consecutive days. One day after the last dose they are killed by cervical dislocation and the uteri removed, freed from adherent tissue, blotted, and the wet or dry weight (to the nearest mg) recorded. Pieces of uterus can be fixed for histological examination if required. Weights of uteri are then compared between groups.

Vaginal opening and cornification of each experimental and control animal should be examined daily and recorded as shown in the results section.

Treatment of Results

Postcoital Antifertility Testing in the Mouse and Rat

As is evident from Table 11.1, oral administration of flavonoids on days 4–6 *post coitum* and days 1–7 *post coitum* proved to be most potent (80%) in preventing nidation. Treatment on days 1–3 *post coitum* resulted in 60% anti-implantational activity, whereas only 40% of anti-implantational activity was recorded at the postimplantational stage of pregnancy, i.e. days 7–9 of pregnancy.

The anticonceptive activity was localized in the petroleum ether-soluble fraction Fr. III. It was observed that at the doses of 10, 20 and 40 mg/rat on days 1–5 *post coitum* the administration of the petroleum ether-soluble

Table 11.1 Post-coital antifertility efficacy of flavonoids of *Grangea maderaspatana* Poir in adult female mice at a dose of 20 mg/kg body weight/day

| Treatment given orally on days of pregnancy (*post coitum*) | Number of animals | Autopsy on day 12 *post coitum* | | Number of females delivered | Number of litters | Inhibition of pregnancy (%) |
		Foetus	Resorption sites			
Control	5	28	2	5	28	0
1–3	5	3	0	2	3	60
4–6	5	1	0	1	1	80
7–9	5	7	11	3	7	40
1–7	5	1	3	1	1	80

From Oestrogenic and pregnancy interceptory efficacy of a flavonoid mixture from *Grangea maderaspatana* Poir (*Artemisia maderaspatana*) in the mouse. Jain S et al (1993), *Phytother. Res.* 7(5), 381–385, with permission.

Table 11.2 Anti-implantation and anticonceptive activity of *Malvaniscus conzattii* extract and its different fractions in adult female rats

Treatment[a] (Fraction)	Dose (mg/rat)	Implantation sites per rat (mean ± SEM)[b]	Pups delivered per rat (mean ± SEM)	Animals with anti-implantation activity (%)	Animals with anticonceptive activity (%)
Control	–	7.3 ± 0.11	6.8 ± 0.10	0	0
M. conzattii					
(90% EtOH)	40	2.6 ± 0.13	2.0 ± 0.27	50.0	60.0
Benzene-soluble	20	1.8 ± 0.23	1.2 ± 0.21	50.0	70.0
(Fr. I)	40	1.2 ± 0.17	0.8 ± 0.17	60.0	80.0
Benzene-insoluble	20	5.0 ± 0.19	4.7 ± 0.17	10.0	10.0
(Fr. II)	40	4.1 ± 0.23	3.3 ± 0.25	20.0	30.0
Petroleum ether-soluble	10	1.5 ± 0.17	1.2 ± 0.17	50.0	60.0
Fr. III)	20	1.0 ± 0.13	0.4 ± 0.08	60.0	80.0
	40	0.5 ± 0.11	NIL	80.0	100.0
Petroleum ether-insoluble	20	5.6 ± 0.30	4.6 ± 0.32	20.0	30.0
(Fr. IV)	40	4.0 ± 0.28	3.0 ± 0.26	30.0	40.0

[a]Days 1–5 of pregnancy; oral; [b]Day 10 of pregnancy. From Estrogenic and anticonceptive activity of petroleum ether soluble fraction of *Malvaviscus conzattii* flowers in female rats. Bhargava S (1987), *Phytother. Res.*, 1(4), 154–157, with permission.

fraction Fr. III resulted in 50%, 60% and 80% anti-implantation, as well as 60%, 80% and 100% anticonceptive activity, respectively (Table 11.2). All the mated animals delivered normally, and none of the pups delivered showed any evidence of teratogenicity up to the age of 1 month.

Testing for Oxytocic Activity

Traces Obtained in vitro   The hot methanol extract (HME) of *Monechma ciliatum* (5×10^{-5}–8×10^{-4} g/ml) produced a concentration dependent contraction of the non-pregnant uteri of the rat, mouse and guinea-pig (Figure 11.3). The extract contracted the oestrus and non-oestrus uteri producing an increase in amplitude and frequency of contraction similar to lower doses of oxytocin (0.00–0.01 IU). The mean ($\pm$SE) maximum tension change in 10 rats induced by HME was 3.2 $\pm$ 0.15 g, while the mean ($\pm$SE) frequency change was 2.0 $\pm$ 0.02/min compared with control (1/min).

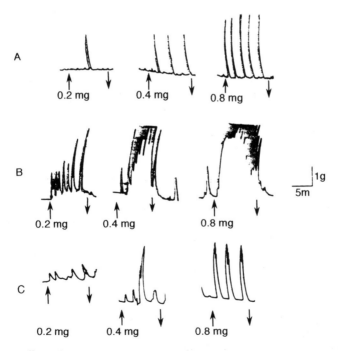

Figure 11.3   Effect of HME (0.2–0.8 mg) on the isolated non-pregnant uteri of mouse (A), guinea-pig (B), and rat (C). ↑ Indicates addition of HME, ↓ indicates washing. From Oxytocic and oestrogenic effects of *Monechma ciliatum* methanol extract *in vivo* and *in vitro* in rodents. Uguru MO *et al* (1995), *Phytother. Res.*, 9(1), 26–29, with permission.

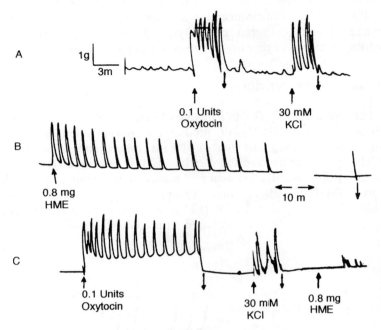

Figure 11.4   Desensitising effect (B) of the extract (0.8 mg) on the non-pregnant iso-lated rat uterus. The experiment also shows the effect of oxytocin and KCl in the absence (A) and presence (C) of HME (0.8 mg). ↑ Indicates addition of oxytocin, KCl or HME, ↓ indicates washing. From Oxytocic and oestrogenic effects of *Monechma ciliatum* methanol extract *in vivo* and *in vitro* in rodents. Uguru MO *et al* (1995), *Phytother. Res.*, **9**(1), 26–29, with permission.

When the rat uterus was exposed to the extract for 15–20 min there was a gradual decline in the responsiveness of the uterus to the extract, such that the contraction was abolished after 60 min (Figure 11.4b). During the period of tachyphylaxis to the extract the rat uterus was not responsive to further doses of the extract (to doses of the HME which initially induced con-tractions). However, after about 30 min of rest and washing at 10 min inter-vals, the rat uterus usually recovered its responsiveness to the extract.

Results in vivo: *Effect of HME on pregnant rat uterus* in situ   The control rats showed a progressive increase in body weight until parturition on day 23 when there was a sudden drop in body weight and all the litters were normal (Figure 11.5). Oral administration of HME (20 mg and 40 mg) did not induce spontaneous abortion compared with oxytocin (Table 11.3). How-ever, laparotomy on day 23 of pregnancy revealed death of the foetuses in all rats that received oral HME. Moreover, there was a significant decrease from control ($p > 0.05$) in the body weight of rats in groups B and C after

Table 11.3   Treatment given to each pregnant rat and the observations on abortions and death *in utero*

Group	Treatment given on days 15, 16, and 17 of pregnancy	Number of immediate abortions	Death *in utero*
A	2 ml of normal saline	0/4	0
B	20 mg of HME per rat	0/4	4
C	40 mg HME per rat	0/4	4
D	2 units of oxytocin per rat	4/4	0

From Oxytocic and oestrogenic effects of *Monechma ciliatum* methanol extract *in vivo* and *in vitro* in rodents. Uguru MO *et al* (1995), *Phytother. Res.*, 9(1), 26–29, with permission.

the administration of the extract (Figure 11.5). In group D, expulsion of live foetuses occurred about 2 h after administration of oxytocin. The litters died about 4 h later.

Testing for Oestrogenicity

Table 11.4 shows that in ovariectomised control animals the vaginal opening remained closed and there was no sign of cornification. When diethylstilboes-

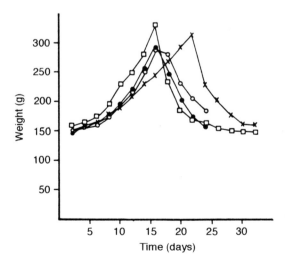

Figure 11.5   Graph of weight against day of pregnancy showing the effect of HME and oxytocin on pregnant rat uterus *in situ*. × group A; ○ group B; ● group C; □ group D, SD range 5.1–9.3. From Oxytocic and oestrogenic effects of *Monechma ciliatum* methanol extract *in vivo* and *in vitro* in rodents. Uguru MO *et al* (1995), *Phytother. Res.*, 9(1), 26–29, with permission.

Table 11.4 Effect of hexane extract of *Ferula jaeschkeana* Vatke on vaginal opening and cornification in bilaterally ovariectomised immature rats

Group No.	Treatment (route)	Dose (mg/kg)	*n*	Vaginal opening (%)	Vaginal cornification (%)
1	Control (vehicle only)	–	10	0	0
2	DES (p.o.)	0.1	8	0	0
		0.5	9	0	0
		0.8	9	11	0
		1.0	8	75*	50*
		1.5	10	80*	80*
		2.0	8	100*	100*
		2.5	9	100*	100*
		3.0	8	100*	100*
3	Hexane extract (p.o.)	5	8	0	0
		10	9	11	0
		15	8	75*	37
		20	9	100*	90*
		25	8	100*	100*
		30	8	100*	100*
		40	8	100*	100*
		50	9	100*	100*

*Statistical significance from control $p < 0.001$.
From Post-coital contraceptive effects of *Ferula jaeschkeana* extracts in rats and hamsters and their hormonal properties. Pathak S and Prakash AO (1989), *Phytother. Res.*, 3(2), 61–66, with permission.

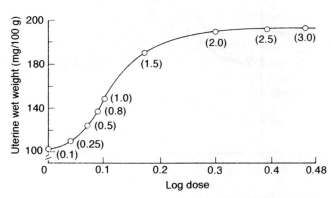

Figure 11.6 Effect of oral administration of diethylstilboestrol on the uterine wet weight of bilaterally ovariectomised immature rats. The dose of mg/kg is shown in parentheses on the curve. From Post-coital contraceptive effects of *Ferula jaeschkeana* extracts in rats and hamsters and their hormonal properties. Pathak S and Prakash AO (1989), *Phytother. Res.*, 3(2), 61–66, with permission.

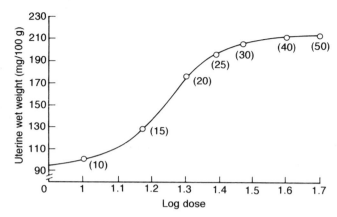

Figure 11.7 Effect of oral administration of hexane extract of *Ferula jaeschkeana* on the uterine wet weight of bilaterally ovariectomised immature rats. The dose in mg/kg is shown in parentheses on the curve. From Post-coital contraceptive effects of *Ferula jaeschkeana* extracts in rats and hamsters and their hormonal properties. Pathak S and Prakash AO (1989), *Phytother. Res.*, 3(2), 61–66, with permission.

terol was administered, the vagina of all the rats showed cornification and opening at a mimimum dose of 2 mg/kg p.o. The administration of the hexane extract also induced vaginal cornification and opening but doses lower than 20 mg/kg did not produce a significant effect. Our results also revealed that when diethylstilboestrol (DES) was administered orally at different doses to ovariectomised immature rats, there was a significant increase in the uterine wet weight when compared to control rats and the response was dose dependent as revealed by a sigmoid curve (Figure 11.6). The administration of hexane extract alone at lower doses (5–10 mg/kg) did not increase the uterine wet weight significantly, but thereafter the uterine weight increased gradually. At 25 mg/kg there was a significant increase and the overall response was also dose-dependent, as revealed through its sigmoid curve (Figure 11.7).

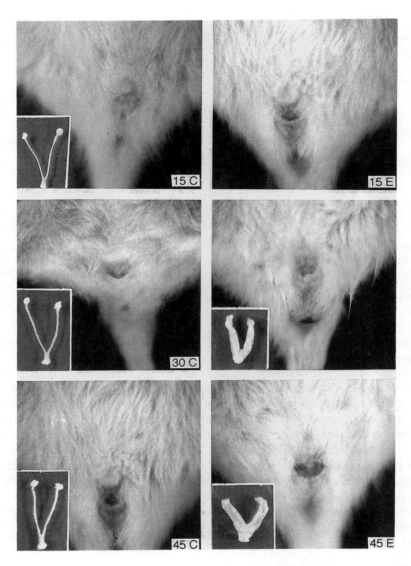

Figure 11.8 Photographs of the ventro-posterior region of the immature rats showing vaginal morphology. Each picture also shows the uterine morphology at the time
of autopsy. For key see Figure 11.9.

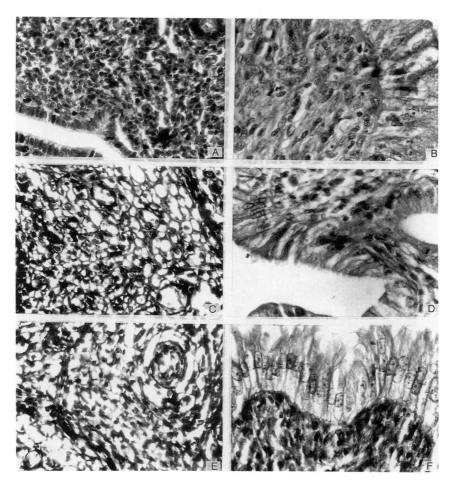

Figure 11.9 (A) Control rat at 15 days schedule. Note that luminal epithelium is composed of single layer of cuboidal cells with enlarged nuclei ($\times$ 400). (B) After 15 days of treatment with hexane extract. Luminal epithelium is elongated and showing leucocytic infiltration ($\times$ 400). (C) Control rat at 30 days of schedule showing loosely arranged stromal cells ($\times$ 400). (D) After 30 days of treatment with hexane extract showing increased height of luminal epithelium and loose stroma ($\times$ 400). (E) Control rats at 45 days schedule showing loose and anastomosed stroma ($\times$ 400). (F) After 45 days of treatment with hexane extract showing hypertrophy of luminal epithelium with finger like projections ($\times$ 400).

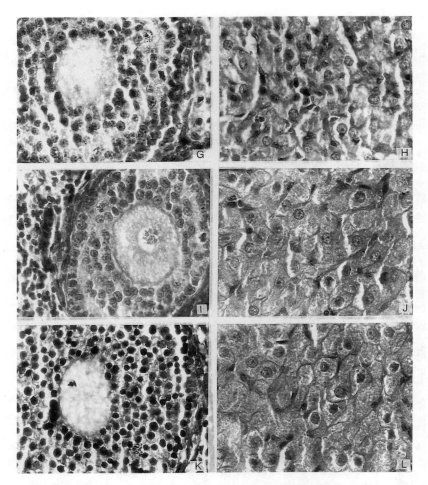

Figure 11.9 (*Continued*).   (G) Control rats at 15 days schedule showing developing follicles with ovum and cells of membrana granulosa are arranged in layers (× 400). (H) After 15 days of treatment with hexane extract. Note that cells within follicle are loosely arranged (× 400). (I) Control rat at 30 days schedule. Note that ovum is surrounded with 3–4 layers of cells of membrana granulosa (× 400). (J) After 30 days treatment with hexane extract. Note the flattened and luteinised cells (× 400). (K) Control rat at 45 days schedule. Mature follicle is having ovum surrounded by darkly stained secretory cells of membrana granulosa (× 400). (L) After 45 days treatment with hexane extract. Note well developed corpus luteum with typical luteinised cells (× 400). From Morphological and functional consequences in the reproductive organs of immature rats exposed to multiple administration of *Ferula Jaeschkeana*. Prakash A *et al* (1996) *Phytother. Res.*, in press.

FURTHER READING

Fertility Control

Reviews

Casey RCD (1960). Alleged antifertility plants of India. *Indian J. Med. Sci.*, **14**, 590–600.

Farnsworth N *et al* (1975). Potential value of plants as sources of new antifertility agents I & II. *J. Pharm. Sci.*, **64**, 535–598 and 717–754.

Karim SSM (1970). Prostaglandins in fertility controls. *Lancet*, **2**, 610–614.

Kato K *et al* (1983). Inhibition of implants and termination of pregnancy in rats by a human chorionic gonadotrophin antagonist. *Endocrinology*, **113**, 195–199.

Morris JM *et al* (1967). Compounds interfering with ovum implantation and development. I. Alkaloids and antimetabolites. *Fert. Ster.*, **18**, 7–17.

Soeijarto DD *et al* (1978). Fertility regulating agents from plants. *WHO Bull.*, **56**, 343–352.

Wu D (1989). An overview of the clinical pharmacology and therapeutic potential of gossypol as a male contraceptive agent and in gynaecological disease. *Drugs*, **38**, 333–341.

Methods

Akah P (1994). Abortifacient activity of some Nigerian plants. *Phytother. Res.*, **8**(2), 106–108.

Che CT *et al* (1984). Studies on *Aristolochia*. III. Isolation and biological evaluation of constituents of *A. indica* roots for fertility regulating activity. *J. Nat. Prod.*, **47**, 331–341.

Prakash AO (1981). Antifertility investigation on embelin—an oral contraceptive of plant origin. Part 1. Biological properties. *Planta Medica*, **41**, 259–266.

Prakash AO *et al* (1992). Anti-implantation mechanism of action of embelin in plants. *Phytother. Res.*, **6**(1), 29–33.

Task Force on Plants for Fertility Regulation (1981). *Bioassay protocols for the Special Programme of Research Development and Training in Human Reproduction.* No. 0045E, 25.11.1981, World Health Organisation, Geneva.

Oxytocic Activity in the Female

Alexandrova M and Soloff MS (1984). Oxytocin receptors and parturition III. Increase in estrogen receptor and oxytocin receptor concentrations in the rat myometrium during prostaglandin $F_{2\alpha}$-induced abortion. *Endocrinology*, **106**, 739–743.

Pettibone DJ *et al* (1991). *In vitro* pharmacological profile of a novel structural class of oxytocin antagonists. *J. Pharmacol. Exp. Ther.*, **256**, 304–308.

Soloff MS *et al* (1979). Oxytocin receptors: triggers for parturition and lactation? *Science*, **204**, 1313–1315.

Methods

Blyth DI (1972). Effect of histamine on the electrical activity of the rat uterus. *Br. J. Pharmacol.*, **45**, 126–128.

Lerner LJ *et al* (1966). Effects of hormone antagonists on morphological and biochemical changes induced by hormonal steroids in immature rat uterus. *Endocrinology*, **78**, 111–124.

Prabhakar E and Nanda Kumar NV (1994). Spasmogenic effect of *Datura metel* root extract on rat uterus and rectum smooth muscles. *Phytother. Res.*, **8**(1), 52–54.

Contraceptive Effects in the Male: Reviews

Anandkumar TC and Waites GMH (Eds) (1985). *Methods for the Regulation of Male Fertility*, ICMR, New Delhi.

Chinoy NJ *et al* (1995). Contraceptive efficacy of *Carica papaya* seed extract in male mice. *Phytother. Res.*, **9**(1), 30–36.

Greep RO (Ed) (1977). *Review of Physiology. Reproductive Physiology II*, Vol. 13, University Park Press, Baltimore.

Hoskins DD *et al* (1978). Initiation of sperm motility in the mammalian epididymis. *Fed. Proc.*, **37**, 2534–2542.

Kamal R *et al* (1993). Efficacy of the steroidal fraction of fenugreek seed extract on fertility of male albino rats. *Phytother. Res.*, **7**(2), 134–138.

Suighal RL and Thomas AJ (1976). *Cellular Mechanisms Modulating Gonadal Action*, Vol 2, University Park Press, Baltimore.

Related Experiments: Progestogenic and Antiprogestogenic Activity

Dohler KD and Wuttke W (1974). Total blockade of phasic pituitary prolactin release in rats. Effect of serum LH and progesterone during estrus cycle and pregnancy. *Endocrinology*, **94**, 1595–1600.

Hensleigh PA and Fainstat T (1979). Corpus luteum dysfunction. Serum progesterone levels in diagnosis and assessment of therapy for recurrent and threatened abortion. *Fertil. Steril.*, **32**, 396–400.

Madjerek ZS (1972). A new bioassay of progestational activity. *Acta Morphol. Neerl. Scand.*, **10**, 259–268.

Ohta Y (1982). Deciduoma formation in rats ovariectomised at different ages. *Biol. Reprod.*, **27**, 308–311.

Pathak S and Prakash AO (1989). Post-coital contraceptive effects of *Ferula jaeschkeana* extracts in rats and hamsters and their hormonal properties. *Phytother. Res.*, **3**(2), 61–66.

Oestrogenic and Anti-oestrogenic Activity

Batra S (1986). Effect of estrogen and progesterone treatment on calcium uptake by the myometrium and smooth muscle of the lower urinary tract. *Eur. J. Pharmacol.*, **127**, 37–42.

Jordan VC *et al* (1978). Nonsteroidal antioestrogens: their biological effects and potential mechanisms of action. *J. Toxicol. Environ. Health*, **4**(2–3), 363–390.

Korach KS (1979). Estrogen action in the mouse uterus; characterization of the cytosol and nuclear receptor systems. *Endocrinology*, **104**, 1324–1332.

Majid E and Senior J (1982). Anti-estrogen modification of uterine responses in the rat. *J. Reprod. Fert.*, **63**, 79–85.

Mukku VR *et al* (1981). Stimulation and inhibitory effects of estrogen and antiestrogen on uterine division. *Endocrinology*, **109**, 1005–1008.

Presentations from the Second International Conference on Phytoestrogens. Little Rock, Arkansas, Oct 17–20, 1993. *Proc. Soc. Exp. Biol. Med.*, **208**(1), 1–138.

Sheehan DM *et al* (1981). Uterine responses to oestradiol in the neonatal rat. *Endocrinology*, **109**, 76–82.

Methods

Bhargava SK (1987). Estrogenic and anticonceptive activity of a petroleum ether-soluble fraction of *Malvaviscus conzattii* flowers in female rats. *Phytother. Res.*, **1**(4), 154–157.

Everett J (1962). The influence of oestradiol and progesterone on the endometrium of the guinea-pig *in vitro*. *Endocrinology*, **24**, 491–496.

Lerner LJ *et al* (1966). Effects of hormone antagonists on morphological and biochemical changes induced by hormonal steroids in immature rat uterus. *Endocrinology*, **78**, 111–117.

Rubin BL *et al* (1971). Bioassay of estrogens using mouse uterine response. *Endocrinology*, **49**, 429–439.

WHO Protocol MB-70, 9856E (1983). A method for detecting estrogenicity in plant extracts administered orally in rats.

Delayed Implanatation Method

Pathak S and Prakash AO (1989). Post-coital contraceptive effects of *Ferula jaeschkeana* extracts in rats and hamsters and their hormonal properties. *Phytother. Res.*, **3**(2), 61–66.

Prakash AO and Roy SK (1984). Induction of implantation by a non-steroidal antifertility agent, 1,2-diethyl 1,3-*bis*(p-methoxyphenyl)-1-propene in rat. *Int. J. Fertility*, **29**(1), 13–15.

Prakash AO *et al* (1992). Anti-implantation mechanism of action of embelin in plants. *Phytother. Res.*, **6**(1), 29–33.

Determination of oestrus

Short DT and Woodnatt DP (1974). *The IAT Manual of Laboratory Animal Practice and Techniques*, p. 340. Granada Publishing, London.

Histological Evaluation of oestrogenicity

Singh P *et al* (1990). Wogonin, 5,7-dihydroxy-8-methoxyflavone as oestrogenic and anti-implantational agent in the rat. *Phytother. Res.* **4**(3), 86–89.

Oestrogenic and Fertility-regulating Substances from Plants

Francis CM *et al* (1967). The distribution of oestrogenic flavonoids in the genus *Trifolium*. *Aust. J. Agric. Res.*, **18**, 47–54.

The Prostate Gland

Reviews

Bauer HW (Ed) (1989) *Benigne Prostatahyperplasie II.* Zuckshwerdt Munich, Wein, San Francisco.
Farnsworth WE (1990). The prostate plasma membrane as an androgen receptor. *Membr. Biochem.*, **9**(2), 141–162.
Farnsworth WE (1991). *The Prostate*, Vol. 19. Wiley–Liss, New York, and refs therein.

Plant Extracts Affecting the Prostate

Adlercreutz H *et al* (1993). Plasma concentrations of phyto-oestrogens in Japanese men. *Lancet*, **342**, 1209–1210.
Belaiche P and Lievoux O (1991). Clinical studies on the palliative treatment of prostatic adenoma with extract of *Urtica dioica*. *Phytother. Res.*, **5**(6), 267–269.
Duker E.-M *et al* (1989). Inhibition of 5α-reductase activity by extracts from *Sabal serrulata*. *Planta Medica*, **55**, 587.
Gansser D and Spiteller G (1995). Aromatase inhibitors from *Urtica dioica* roots. *Planta Medica*, **61**, 138–140.
Hirano T *et al* (1994). Effects of stinging nettle root extracts and their steroidal components on the Na^+, K^+-ATPase of benign prostatic hyperplasia. *Planta Medica*, **60**, 30–33.
Sultan C *et al* (1984). Inhibition of androgen metabolism and binding by a liposterolic extract of *Serenoa repens* in human foreskin fibroblasts. *J. Steroid Biochem.*, **20**(1), 515–519.

Methods for Measurement of 5α-Reductase and Receptor Binding

Breiner M *et al* (1986). Inhibition of androgen receptor binding by natural and synthetic steroids in cultures human genital skin fibroblasts. *Klin. Wochenschr.*, **64**, 732–737.
Hirano T *et al* (1989). Effects of synthetic and naturally occurring flavonoids on Na^+, K^+-ATPase: aspects of the structure–activity relationship and action mechanism. *Life Sci.*, **45**(12), 1111–1117.
Leshin M *et al* (1978). Hereditary male pseudohermaphroditism associated with an unstable form of 5-α-reductase. *J. Clin. Invest.*, **62**, 685–691.

Appendix I: List of Standard Textbooks

Pharmacognosy

Evans WC (1983). *Textbook of Pharmacognosy*, 12th Edn, Baillière Tindall, Eastbourne.
Harborne JB and Baxter H (Eds) (1993). *Phytochemical Dictionary: A Handbook of Bioactive Compounds from Plants*, Taylor and Francis, London.
Williamson EM and Evans FJ (1988). *Potters New Cyclopaedia of Botanical Drugs and Preparations.* CW Daniels, Saffron Walden.

Practical (Experimental) Pharmacology

The Staff of the Department of Pharmacology, University of Edinburgh. (1970). *Pharmacological Experiments on Intact Preparations*, 2nd Edn. E and S Livingtone, Edinburgh.
The Staff of the Department of Pharmacology, University of Edinburgh. (1970). *Pharmacological Experiments in Isolated Preparations*, 2nd Edn. E and S Livingtone, Edinburgh.
Turner RA (1965). *Screening Methods in Pharmacology*, Academic Press, New York.

Theoretical Pharmacology

Gilman AG, Rall TW, Neis AS and Taylor P (1990). *The Pharmacological Basis of Therapeutics*, 8th Edn, Pergamon, Oxford.
Katzung BG (1995). *Basic and Clinical Pharmacology*, 6th Edn, Appleton and Lange, Connecticut.
Okpako DT (1991). *Principles of Pharmacology: A Tropical Approach*, Cambridge University Press, Cambridge.
Rang HP, Dale MM and Ritter JM (1995). *Pharmacology*, 3rd Edn, Churchill Livingstone, Edinburgh.

Statistics and Calculations

Swinscow TDW (Ed) (1985). *Statistics at Square One*, 8th Edn, British Medical Association, London.
Tallarida RJ and Murray RB (1986). *Manual of Pharmacologic Calculations with Computer Programs*, 2nd Edn, Springer Verlag, New York.

Appendix II: Physiological Saline Solutions (PSS) Used for Bathing Isolated Tissues

Concentrations are always expressed as millimolar (mM), although formulae for making up are given in grams per litre (g/l) for convenience. Minor variations on these formulae are often used by different research groups. Modified Tyrode or Krebs solutions may be used, for example 'calcium-free' (if calcium channel activity is under investigation, this may include the addition of EGTA, 1 mM).

Tyrode Solution

For duodenum, ileum and other preparations. Maintained at $37\,^{\circ}C$ and aerated with 95% O_2, 5% CO_2.

Salt	Conc (mM)	Conc (g/l)
NaCl	137.0	8.0
KCl	2.6	0.2
$CaCl_2$	0.3	1.8
$NaHCO_3$	11.9	1.0
$MgCl_2. 6H_2O$	1.05	2.0
$NaH_2PO_2. 2H_2O$	0.4	0.062
Glucose	5.5	1.0

'Double Glucose' Tyrode Solution

For rat phrenic nerve–diaphragm, rabbit aorta, etc. Maintained at $37\,^{\circ}C$ and aerated with 95% O_2, 5% CO_2.

As above, although some workers omit the sodium acid phosphate and double the concentration of potassium chloride, as well as that of glucose. The magnesium concentration is sometimes increased.

219

Krebs Solution

For rat fundic strip and isolated phrenic nerve–diaphragm. Maintained at 37 °C and aerated with 95% O_2, 5% CO_2.

Salt	Conc (mM)	Conc (g/l)
NaCl	115.0	6.8
KCl	4.7	0.35
$CaCl_2$	2.5	0.28
$NaHCO_3$	25.0	2.1
$MgCl_2. 6H_2O$	1.05	0.11
$NaH_2PO_2. 2H_2O$	1.2	0.18
Glucose	11.0	2.0

Krebs–Hensleit Solution

For rat and guinea-pig isolated atria, rabbit thoracic aorta and myocardial strips, and similar. Maintained at 37 °C and aerated with 95% O_2, 5% CO_2.

Salt	Conc (mM)	Conc (g/l)
NaCl	115.0	6.8
KCl	4.7	0.35
$CaCl_2$	2.5	0.28
$NaHCO_3$	25.0	2.1
$MgSO_4. 7H_2O$	1.5	0.25
$K_2H_2PO_4. 2H_2O$	1.2	0.16
Glucose	11.0	2.0

De Jalon (De Jalon–Ringer) Solution

This usually contains lower calcium and half the usual glucose concentrations, to reduce spontaneous contractions in the uterine horn or uterine strip preparations for which it is used. Maintained at 37 °C or slightly lower and aerated with 95% O_2, 5% CO_2.

Salt	Conc (mM)	Conc (g/l)
NaCl	154.0	9.0
KCl	5.5	0.42
$CaCl_2$	0.01	0.06
$NaHCO_3$	6.00	0.5
Glucose	2.75	0.5

Index

Index compiled by C. Purton